Pre-Algebra

Second Edition

Ideal for Independent practice!

Achieve a solid foundation in Algebra in a very short time!

Contents

Welcome Students!

You did the right choice in deciding to work with this guide. Your time is very valuable and I don't want you to waste it in long exercises and tedious explanations that do not add any meaningful understanding to the concepts and skills required to master pre-algebra.

Instead, this guide is just straight to the point! Each and almost every page starts with quick and short example or explanation immediately followed by the skill practice. Work on these exercises and check the selected answers (usually odd numbers) at the back of the book! That is it. In no time, you master the pre-algebra concepts and be ready to cruise algebra at high speed!

Congratulations in deciding to master the pre-algebra which is the true entrance to advanced algebra!

Please do not hesitate to write me any comments or questions about this book at aalibotan@gmail.com.

1)

$$398 \times 72$$

$$305 \times 99$$

$$8\overline{)128}$$

$$62\overline{)6918}$$

2)

$$1\frac{5}{9} + 1\frac{2}{5}$$

$$7\frac{3}{5} - \frac{2}{7}$$

$$2\frac{2}{9} \times 2\frac{7}{10}$$

$$1\frac{1}{5} \div 3\frac{1}{4}$$

3)

$$11 - 0.15$$

$$5.24 \times 5.8$$

$$4.2\overline{)84}$$

$$0.06\overline{)3}$$

4)

Fraction	Decimal	Percent
$\frac{3}{15}$	?	?
?	0.04	?
?	?	2.5%

5) 9 ft 4 in + 7 ft 9 in.

6) The cost of a book was $12 before adding 8% tax. What is the total cost?

7) Calculate the diameter, the circumference and the area of a circle with a radius of 12 cm.

1) Integers: Absolute Values

Integers: Include whole numbers and negative numbers.

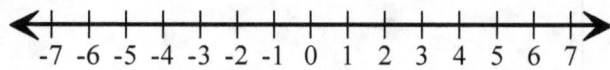

- **Gaining or more of something** is positive on number line.
- **Less of something is negative** in the number line.
- **The absolute value** of any number is always positive : $|-9| = 9$
- **For two negative numbers**, the one closer to the zero is greater: $-2 > -7$

Is the integer in each situation positive or negative?

1. The temperature of Minneapolis is 20 degrees below zero.

2. Amina has lost seventy five dollars.

3. Farah has gained a profit of $300.

4. Omar has lost 20 pounds before gaining 30 pounds.

5. A baby grows 9 inches taller.

6. Ali rides the hill for 200 feet.

7. Sofia has removed 10 books from the store.

Evaluate the expressions

8) $|-20|$

9) $|-12|$

10) $|-12| + |-13|$

11) $|-15| + |2|$

12) $-|-17|$

13) $|-12| + |10|$

14) $|-13| + |-6| - 5$

15) $28 - |-17|$

16) $|-12| + |10| - 10$

Extra Practice: Evaluate the expressions

1) $\lvert -2 \rvert + \left\lvert \dfrac{-20}{4} \right\rvert$	2) $\lvert -12 \rvert + \dfrac{\lvert -8 \rvert}{\lvert -4 \rvert}$	3) $\lvert -1 \rvert - \lvert -1 \rvert$
4) $3 \times \lvert -10 \rvert + \lvert -5 \rvert$	5) $10 - \lvert -9 \rvert +$	6) $\lvert -14 \rvert + \lvert 8 \rvert$
7) $\lvert -3 \rvert \times 3 + \lvert -6 \rvert \times 2$	8) $20 + \lvert -1 \rvert \times \lvert -2 \rvert$	9) $\left\lvert \dfrac{-10}{2} \right\rvert + \lvert 10 \rvert$
10) $\lvert 8 \rvert + \left\lvert \dfrac{-10}{2} \right\rvert$	11) $\lvert -12 \rvert + \left\lvert \dfrac{-12}{4} \right\rvert \times 4$	12) $\lvert -12 \rvert + \lvert -12 \rvert$
13) $\lvert -15 \rvert + \lvert 0 \rvert$	14) $18 - \lvert -17 \rvert$	15) $\lvert -10 \rvert + \lvert -10 \rvert -$
16) $\lvert -1 \rvert + \lvert -6 \rvert - 5$	17) $9 + \lvert -1 \rvert \times \lvert -5 \rvert$	18) $\lvert -12 \rvert + \lvert 10 \rvert - 10$

2) Ordering Integers: Which is bigger?

-5 -4 -3 -2 -1 0 1 2 3 4 5

- A **positive number** is always greater than negative number: $4 > -10$
- For two **negative numbers**, the one closer to the zero is greater: $-2 > -5$
- The **absolute value** of any number is always positive: $|-7| = 7$

A) Use either a $<$ or $>$ to say the truth about a number:

1) –8 $\bigcirc$ –3

2) –3 $\bigcirc$ –9

3) 0.5 $\bigcirc$ –2

4) $|-7|$ $\bigcirc$ 20

5) –2 $\bigcirc$ –1/2

6) 0 $\bigcirc$ –5

7) –17 $\bigcirc$ –19

8) $|-6|$ $\bigcirc$ –24

9) –3/4 $\bigcirc$ –1/2

10) –4.5 $\bigcirc$ –5.05

11) 2.5 $\bigcirc$ –7.05

12) $|-12|$ $\bigcirc$ 11

B) Order these numbers from smallest to largest

1) -7, -5, 5 , -9, 0, $|$-10 $|$

2) -4, -11, 15 , -91, 0 , $|-15|$

3) $|-4|$, -4 , 0, -11, 13, $|-14|$

4) $|-14|$, -14 , 10, -12, 1, $|-11|$

5) -1, -5, $|-7|$, -4 , -10, -11

6) $|-9|$, -9 , -10, -12, -21

11

3) Add Integers-A

- **To add integers of the same signs:** add numbers and keep their sign.
    - **1)** $(-2) + (-6) = -8$ **2)** $-5 + -6 = -11$ **3)** $(5) + (4) = 9$

- **To add integers of the different signs:** Just subtract and choose the sign of the larger number (in absolute value):
    - **1)** $(2) + (-6) = -4$ **2)** $-8 + 3 = -5$ **3)** $-5 + 9 = 4$

1) $-6 + (-4)$ 11. $9 + (-14)$

2) $-9 + (-2)$ 12. $-14 + 9$

3) $8 + (-4)$ 13. $-33 + (-4)$

4) $-1 + 5$ 14. $-21 + (10)$

5) $-13 + 4$ 15. $-17 + (-4)$

6) $-13 + (-4)$ 16. $-16 + (5)$

7) $-15 + 21$ 17. $-20 + (-4)$

8) $10 + (-14)$ 18. $-13 + 22$

9) $8 + (-4)$ 19. $40 - 35$

10) $14 + (-12)$ 20. $-13 + (-4)$

4) Add Integers-B

1) $-3 + (-10)$

2) $-15 + (-6)$

3) $15 + (-4)$

4) $-15 + 5$

5) $-1 + 14$

6) $28 + (-4)$

7) $-5 + (-21)$

8) $17 + (-4)$

9) $-12 + (-24)$

10) $-26 + (-10)$

11) $-14 + 4 + (-5)$

12) $20 + (-14) + (-5)$

13) $-5 + (-5) + (-7)$

14) $-10 + 7 + (-5)$

15) $-20 + 6 + (-3)$

16) $-10 + 10 + (-8)$

17) $26 + (-26) + (-8)$

18) $0 + (-14) + (-5)$

19) $12 + (-10) + (-5)$

20) $6 + (-4) + (-5)$

5) Subtract Integers A

1) Connect **any two consecutive negatives as one positive, then simplify.**	2) **Change** any consecutive positive and negatives into negative, then simplify.
$6 - (-4) = 6 + 4 = 10$ $12 - (-5) = 12 + 5 = 17$	$-9 - (+4) = -9 - 4 = -13$ $12 - (+4) = 12 - 4 = 8$

1. $-7 - (-3)$

2. $-9 - (-2)$

3. $8 - (-4)$

4. $-1 - 5$

5. $-13 - 4$

6. $-13 + (-4)$

7. $-15 - 21$

8. $-13 - (-4$

9. $8 - (-4)$

10. $9 - 14$

11. $9 - (-14)$

12. $-14 - 9$

13. $-33 - (-4)$

14. $-21 - (10)$

15. $-17 - (-4)$

16. $-16 - (3)$

17. $-10 - (-4)$

18. $13 - (-2)$

19. $-40 - 35$

20. $-13 - (-4)$

6) Subtract Integers B

1. $8 - (-9)$

2. $19 - (-6)$

3. $-14 - 24$

4. $-10 - (-5)$

5. $-20 - 4$

6. $-10 + (-14)$

7. $20 - 41$

8. $-20 - (-10)$

9. $65 - (-25)$

10. $-40 - 50$

11. $-10 - 4 - (-5)$

12. $-8 + (-14) - (-5)$

13. $-5 - (-5) + 7$

14. $11 + 4 - (-12)$

15. $-30 - 6 - (-36)$

16. $-10 - 10 - (-8)$

17. $-23 + (-26) - 8$

18. $8 - (-14) + (-5)$

19. $-12 - (-10) + (-5)$

20. $5 - (-6) - (-5)$

7) Multiply Signed Integers

> ✓ **To multiply same sign** : Multiply the numbers and keep your answer positive:
>   *Examples:* 1) $-6 \times -4 = 24$ 2) $(-3)(-5) = 15$ 3) $4 \times 5 = 20$

> ✓ **To multiply different signs:** Multiply numbers & keep your answer negative:
>   *Examples:* 1) $-7 \times 3 = -21$ 2) $(5)(-2) = -10$ 3) $5(-7) = -35$

1. $-3 \times (-4)$

2. $(-9)(-2)$

3. $8 \times (-4)$

4. -1×5

5. -13×4

6. $-13 \times (-4)$

7. -15×-2

8. $10 \times (-2)$

9. $-1 \times (-4)$

10. 9×-4

11. $9 \times (-1)$

12. -1×-9

13. $(-3)(-4)$

14. $-11 \times (3)$

15. -7×-4

16. $-16(5)$

17. $-8 \times 8 \times 5$

18. $-3 \times -4 \times -5$

19. $-1 \times -2 \times -5$

20. $-10 \times 2 \times -5$

8) Divide Signed Integers

To divide same sign integers : Divide the numbers and keep your **answer positive:**	
Examples: 1) $-6 \div -2 = 3$ 2) $(-20) \div (-5) = 4$ 3) $14 \div 2 = 7$	
To divide different sign integers: Divide numbers and keep the **answer negative:**	
Examples: 1) $-16 \div 2 = -8$ 2) $(15) \div (-3) = -5$ 3) $35 \div (-7) = -35$	

1. $-8 \div -4$

2. $-20 \div 4$

3. $8 \div (-4)$

4. $(-30) \div (-3)$

5. $(-9) \div (-3)$

6. $-12 \div (-4)$

7. $24 \div -4$

8. $-10 \div 10$

9. $-4 \div (-4)$

10. $-18 \div -2$

11. $9 \div (-1)$

12. $-7 \div -1$

13. $-5 \div 1$

14. $10 \div (-2)$

15. $-9 \div -1$

16. $-16 \div -4$

17. $-4 \div -2$

18. $-8 \div 8 \times 5$

19. $-20 \div -4 \times -5$

20. $7 - 12 \div (-4)$

9) Add and Subtract Algebraic Expressions

✓ Combine like terms only: ***(combine only x and x and not x and y)***

✓ Remember also: it is usual to write (x) rather than (1x)

Examples: 1) $4x - 5x = (4 - 5)x = -x$ 2) $-5x + 5x = (5 - 5)x = 0$

3) $5x - 5y = 5x - 5y$ 4) $8y - 6y = (8 - 6)y = 2y$

1.   **−6x − 4x**   11.   **19y − 14y**

2. $-3y - 2y$ 12. $-w - 9w$

3. $8x - (-4y)$ 13. $-3t - 14t$

4. $-x - 3x$ 14. $-20x - x$

5. $-2a - 4b$ 15. $-17t - 4t$

6. $3y - (-4y)$ 16. $-16x - (5)$

7. $-13m - 4m$ 17. $-8x + 8x - 5x$

8. $-10x - (-4x)$ 18. $6y - 20y + 5y$

9. $-8x - (9)$ 19. $-12y - (-5y)$

10. $6x - (-4x)$ 20. $25k - (-4k) + 3y$

10) Multiply & Divide Simple Algebraic Expressions

1. $(+) \times (+) = (+)$	Example 1: $3 \times 2 = 6$
2. $(-) \times (-) = (+)$	Example 2: $-3y \times -5 = 15y$
3. $(-) \times (+) = (-)$	Example 3: $-3y \times 5 = -15y$
4. $(+) \times (-) = (-)$	Example 4: $4y(-5x) = -20xy$
1. $(+) \div (+) = (+)$	Example 1: $10 \div 2 = 5$
2. $(-) \div (-) = (+)$	Example 2: $-10x \div -5 = 2x$
3. $(-) \div (+) = (-)$	Example 3: $-15x \div 5 = -3x$
4. $(+) \div (-) = (-)$	Example 4: $15x \div -5 = -3x$

Simplify Each Expression: **Remember** *the big dot* ($\bullet$) and the parenthesis ( ) mean times.

1. $-6 \bullet (-4)$

2. $6x \div (-3)$

3. $-3z \times -4y$

4. $-4r(8s)$

5. $-2a(-4b)$

6. $9y \times (-3)$

7. $-3(2) \times (-4xy)$

8. $-4r \times (8s)$

9. $-3(x)(-y)$

10. $-2(-4c)$

11. $-6 \bullet (-4x)$

12. $-4x \bullet (-4)$

13. $-4z(0)(xy)$

14. $(-8)(-2)(-3x)$

15. $(-5x)(-2)(3)$

16. $-9 \bullet (-6y)$

17. $-4(-4w)$

18. $-6y\,(-4)(-2)$

19. $-3w(-4)(2)$

20. $3(-2)(-2w)$

11) Evaluate Algebraic Expression I

Examples: **Evaluate Expressions if x= -2, y =3**		
1) $x - 5 =$ $-2 - 5 = -7$	**2)** $-6 - x$ $= -6 - (-2)$ $= -6 + 2 = -4$	**3)** $2x - 2y + b =$ $2(-2) - 2(3) =$ $-4 - 6 = -10$

Directions: Evaluate Expressions if x= -2, y =3, a= -3 and b= 4

1. $6x - 4x$

2. $-3y - 2y$

3. $3x + 4y$

4. $-3x + 3x$

5. $-2a - 4b$

6. $3y - (-4y)$

7. $13x - 4x$

8. $-6x - (-4y)$

9. $-8x - (9)$

10. $6x - (-4x)$

11. $19y - 14y$

12. $-2a - 6b$

13. $-3x - 14y$

14. $14y - x$

15. $-7b - 4a$

16. $-16x - (5)$

17. $-4x + 6y - 2x$

18. $6y - 2y - 5x$

19. $-12y - (-5y)$

20. $5a - (-4b) + 3y$

12) Evaluate Algebraic Expressions II

Example: Evaluate Expression if x= -2 and y =3

a) $-6x\,(5y) =$	**b)** $(2)(-4xy)=$
$-6(-2)(5\cdot 1) = 12(5)= 60$	$(2)\,(-4\cdot -2\cdot 3)= 2(8\cdot 3)= 2(24) =48$

Evaluate Each expression if $x= 3$, $y=-1$, $a= 2$ and $b =-2$

1) $-8\cdot(-4x)$

2) $9y\cdot(-3)$

3) $-6\cdot(-4x)$

4) $-9\cdot(-6y)$

5) $6x\div(-3)$

6) $-3(2)(-4xy)$

7) $-4x\cdot(-4)$

8) $-4b(-4a)$

9) $-3a\cdot -4y$

10) $-4a(8x)$

11) $-4xy$

12) $-6y\,(-4)$

13) $-4(8y)$

14) $-3(x)(-y)$

15) $(-8x)(-3x)$

16) $-3a(-4)$

17) $-2a(-4b)$

18) $-2a(-4b)$

19) $(-5x)(-2a)$

20) $(-2x)(-2a)$

13) Evaluate Algebraic Expressions III

> To evaluate variables, replace the variables with the numbers and simplify:
>
> *Example:* **Evaluate** $x(y+z)$ *when x = 5, y = 3 and Z = 1*
>
> $5(3+1) = 5 \times 4 = 20$

Evaluate each expression if x = 6, y = 3, and z = 2.

1. $x + 2y + z$　　　　　　　　　**2.** $3x - y$

3. $x + y - z$　　　　　　　　　　**4.** $3x - y + 3z$

5. $12z - x$　　　　　　　　　　　**6.** $3(x + y + z)$

7. $xy \div z$　　　　　　　　　　　**8.** $xyz - x$

Evaluate each expression if a = 8, b = 4, c = - 6, and d = - 3.

9. $a + b - c$　　　　　　　　　　**10.** $a + b - (c + d)$

11. $3a + 4d$　　　　　　　　　　**12.** $bc - d$

13. $(a + b) \div (c - d)$　　　　　**14.** $c(4 + d)$

15. $ab - cd$　　　　　　　　　　**16.** $bc + a - d$

17. Suuban's age is two years more than twice of her brother's age.

   a. Write an expression for Suuban's Age.

   b. If Suuban's brother is currently 6 years, how old is Asha?

18. The total math grade for Mark, Abdi and Layla is 245.

   a. Using variables x, y and z, write an expression for the three students' score.

   b. If Mark and Abdi scored 170, what is the score for Layla?

14) Equations: Find the Solution

Example: Which number is the solution of the equation? $x + 15 = 21$; **4, 5, or 6?**

The Solution is **6.** Why? Just substitute **x** for **6** to have

$$6 + 15 = 21 \text{ OR } 21 = 21$$

Note: Other numbers such as **4** and **5** wouldn't work. Try them!

1) $x + 15 = 35$; 7, 8, 20

2) $x - 15 = 5$; 10, 15, 20

3) $x + 3 = 25$; 21, 22, 3

4) $x - 7 = 14$; 28, 25, 21

5) $4 + x = 11$; 7, 5, 6

6) $7 + x = 13$; 6, 7, 8

7) $18 - x = 3$; 14, 15, 16

8) $21 - x = 7$; 14, 15, 18

Find the Solution mentally:

9) $x + 20 = 21$

10) $x - 10 = 21$

11) $x + 17 = 22$

12) $x - 15 = 2$

13) $14 - x = 2$

14) $27 - x = 12$

15) $3 - x = 0$

16) $7 - x = 2$

15) Solve One Step Equations: Addition

Follow the examples to solve for x, y, or z:

1) $x + 15 = 35$
Isolate the x by subtracting 15 from both sides: $$x + 15 - (15) = 35 - (15)$$ $$x = 20$$

2) $x + 3 = 12$
Isolate the x by subtracting 3 from both sides: $$x + 3 - (3) = 12 - (3)$$ $$x = 9$$

1) $x + 4 = 10$ 2) $x + 9 = 12$ 3) $x + 2 = -1$ 4) $x + 7 = -4$

5) $x + 3 = 10$ 6) $y + 8 = 0$ 7) $x + 16 = 10$ 8) $x + 9 = 1$

9) $x + 14 = 14$ 10) $x + 3 = 11$ 11) $x + 8 = -1$ 12) $y + 9 = 11$

13) $x + 7 = 11$ 14) $z + 11 = 4$ 15) $x + 8 = -12$ 16) $x + 9 = -5$

17) $y + 4 - 3 = 11$ 18) $x + 4 + 2 = 10$ 19) $z + 1 + 2 = 12$ 20) $x + 9 = 0$

16) Solve One Step Equations: Subtraction

Follow the examples to solve for x, y, or z:

1) $x - 9 = 10$
Isolate the x by adding 9 to both sides: $$x - 9 + (9) = 10 + (9)$$ $$x = 19$$

2) $x - 5 = -5$
Isolate the x by adding 5 to both sides: $$x - 5 + (5) = -5 + (5)$$ $$x = 0$$

1) $x - 4 = 10$ 2) $x - 2 = -10$ 3) $x - 3 = -1$ 4) $x - 7 = -4$

5) $x - 3 = 12$ 6) $y - 9 = 0$ 7) $x - 6 = -10$ 8) $x - 9 = -9$

9) $x - 12 = 10$ 10) $x - 3 = 10$ 11) $x - 8 = -1$ 12) $y - 9 = 11$

13) $x - 7 = -11$ 14) $z + 1 = -4$ 15) $x - 8 = -2$ 16) $x - 2 = -5$

17) $y - 3 - 6 = 11$ 18) $x + 1 - 2 = -3$ 19) $z - 1 - 2 = -12$ 20) $x - 8 = 0$

17) One Step Equations: Addition & Subtraction

Follow the examples to solve for x, y, or z:

1) $-9 + x = -1$
Isolate the x by adding 9 to both sides: $$x - 9 + (9) = -1 + (9)$$ $$x = 8$$

2) $-5 = 8 + x$
Isolate the x by subtracting 8 from both sides: $$-5 - (8) = 8 - 8 + x$$ $$-13 = x$$

1) $x + 5 = 1$

2) $x - 3 = -6$

3) $3 + x = -7$

4) $4 - 7 = x$

5) $-9 + y = -2$

6) $y + 9 = 6$

7) $y - 6 = -10$

8) $2 + x = -14$

9) $-x - 8 = 4$

10) $-13 = x + 4$

11) $x - 8 = -1$

12) $y - 9 = 11$

13) $x - 7 = -11$

14) $z + 1 = -4$

15) $x + 8 = -12$

16) $x - 3 = -8$

17) $12 + x - 6 = 10$

18) $x - 2 + 5 = -10$

19) $z + 8 - 10 = -1$

20) $x - 15 = 0$

18) Solve One Step Equations: Multiplication

See these four examples:

1. $8x = 16$	$\dfrac{8x}{8} = \dfrac{16}{8}$	$x = 2$	
2. $5x = -20$	$\dfrac{5x}{5} = \dfrac{-20}{5}$	$x = -4$	

3) $-3x = -1$	$\dfrac{-3x}{-3} = \dfrac{-1}{-3}$	$x = \dfrac{1}{3}$	
4) $-x = 10$	$\dfrac{-x}{-1} = \dfrac{10}{-1}$	$x = -10$	

1) $2x = 10$ **2)** $3x = 9$ **3)** $2x = -10$ **4)** $4x = -4$

5) $-2x = -12$ **6)** $-6y = -18$ **7)** $-3y = -24$ **8)** $-7x = -21$

9) $-4x = -14$ **10)** $-3x = -1$ **11)** $-5x = 10$ **12)** $2y = -14$

13) $8x = -16$ **14)** $-z = 4$ **15)** $-x = -1$ **16)** $-x = 5$

17) $-4y = 36$ **18)** $-x = 0$ **19)** $2y = -1$ **20)** $-32y = -8$

19) Solve One Step Equations: Division

Example 1:	$\dfrac{x}{2} = 12$	$x = 2 \times 12$	$x = 24$
Example 2:	$\dfrac{x}{5} = -3$	$x = 5 \times -3$	$x = -15$
Example 3:	$\dfrac{x}{-4} = -9$	$x = -4 \times -9$	$x = 36$
Example 4:	$\dfrac{-x}{2} = 10$	$-x = 2 \times 10$	$x = -20$

1) $\dfrac{x}{5} = 2$

2) $\dfrac{x}{4} = -2$

3) $\dfrac{x}{8} = -12$

4) $\dfrac{x}{-5} = 7$

5) $\dfrac{x}{-4} = -9$

6) $\dfrac{x}{-6} = -2$

7) $\dfrac{x}{-3} = -12$

8) $\dfrac{x}{-7} = -7$

9) $\dfrac{-x}{7} = -2$

10) $\dfrac{x}{-9} = -7$

11) $\dfrac{-x}{8} = -2$

12) $\dfrac{x}{-12} = 2$

13) $\dfrac{x}{-16} = 2$

14) $\dfrac{x}{-8} = 0$

15) $\dfrac{x}{15} = -8$

16) $\dfrac{x}{9} = -8$

17) $\dfrac{x}{5} = 2$

18) $\dfrac{-x}{3} = 11$

19) $\dfrac{x}{-25} = 0$

20) $\dfrac{x}{-2} = 10$

20) One Step Multiplication/Division Equations

Follow the examples to solve for x, y, or z:

---

1) $15 = -5x$

Isolate the x by dividing -5 to both sides
$$\frac{15}{-5} = \frac{-5x}{-5} \quad or \quad -3 = x$$

---

2) $\frac{x}{4} = 8$

Isolate the x by multiplying 4 to both sides:
$$\frac{4 \times x}{4} = 4 \times 8 \quad or \quad x = 32$$

---

1) $4x = 16$

2) $3x = -1$

3) $\frac{x}{-2} = -7$

4) $-5x = 15$

5) $\frac{x}{9} = -1$

6) $x + 3 = -27$

7) $-x = -1$

8) $-5 = \frac{x}{5}$

9) $-11 = \frac{x}{-5}$

10) $20 = y - 8$

11) $\frac{x}{-7} = 0$

12) $-y = 11$

13) $-\frac{x}{8} = 2$

14) $20x = -5$

15) $2x = -10$

16) $\frac{x}{-6} = -9$

17) $-x = -51$

18) $27 = -3y$

19) $30x = -5$

20) $-7x = -5$

21) Solve One Step Equations: Fractions

Follow the examples to solve for x, y, or z:

1) $x + \dfrac{1}{7} = \dfrac{5}{7}$

Isolate the x by subtracting $\dfrac{1}{7}$ from both sides and then simplify the fraction as needed:

$$x + \dfrac{1}{7} - \dfrac{1}{7} = \dfrac{5}{7} - \dfrac{1}{7} \quad or \; x = \dfrac{5-1}{7} = \dfrac{4}{7}$$

2) $x - \dfrac{1}{2} = \dfrac{3}{8}$

Isolate the x by adding $-\dfrac{1}{2}$ to both sides and then simplify the fractions:

$$x - \dfrac{1}{2} + \dfrac{1}{2} = \dfrac{3}{8} + \dfrac{1}{2} \quad or \; x = \dfrac{3}{8} + \dfrac{1}{2}$$

$$x = \dfrac{3}{8} + \dfrac{1 \times 4}{2 \times 4} \quad or \; x = \dfrac{7}{8}$$

1) $x + \dfrac{2}{5} = \dfrac{3}{5}$

2) $x + \dfrac{2}{9} = \dfrac{5}{9}$

3) $x + \dfrac{1}{7} = \dfrac{5}{7}$

4) $x - \dfrac{1}{5} = \dfrac{3}{10}$

5) $x - \dfrac{5}{6} = \dfrac{1}{2}$

6) $x + \dfrac{1}{2} = \dfrac{3}{5}$

7) $x + \dfrac{3}{8} = \dfrac{1}{5}$

8) $x + \dfrac{3}{7} = \dfrac{11}{21}$

9) $4 = x - \dfrac{3}{7}$

10) $\dfrac{1}{3} = x - \dfrac{3}{5}$

11) $x + \dfrac{1}{4} = 0$

12) $x + \dfrac{1}{3} = \dfrac{3}{7}$

22) Solve One Step Equations: Decimals

Follow the examples to solve for x, y, or z:

1) $15 = 0.5x$
Isolate the x by dividing 0.5 to both sides $\dfrac{15}{0.5} = \dfrac{0.5x}{0.5}$ *or* $30 = x$

2) $x - 1.7 = 8.3$
Isolate the x by adding 1.7 from both sides: $x - 1.7 + 1.7 = 8.3 + 1.7$ *or* $x = 10$

1) $2x = 1.6$

2) $0.33x = 0.66$

3) $\dfrac{x}{-2} = 0.7$

4) $x - 1.5 = 3$

5) $\dfrac{x}{0.2} = 2.5$

6) $x + 3.7 = -2.8$

7) $0.1x = -1$

8) $-0.2 = \dfrac{x}{5}$

9) $-10 = \dfrac{x}{1.2}$

10) $10.5 = y - 2.8$

11) $\dfrac{x}{-7} = 0.1$

12) $-y = 1.9$

13) $\dfrac{x}{0.4} = -0.25$

14) $-2.5x = -12.5$

15) $0.02x = -10$

16) $\dfrac{x}{6} = 0.5$

17) $0.3x = -5.1$

18) $2.7 = 3y$

19) $x + 12.5 = -7.2$

20) $-7x = 3.5$

23) Solve One Step Equations: Mixed Review

1) $-3x = 21$

2) $x + 3 = -10$

3) $\dfrac{x}{-3} = -6$

4) $-5x = 5$

5) $\dfrac{x}{8} = -12$

6) $3x = -30$

7) $2 = -10x$

8) $\dfrac{x}{5} = 2$

9) $-13 = \dfrac{x}{5}$

10) $2 = y + 3$

11) $\dfrac{x}{-2} = 0$

12) $-y = -17$

13) $\dfrac{x}{4} = -6$

14) $20 + x = 4$

15) $x + \dfrac{1}{6} = \dfrac{1}{3}$

16) $-2 = \dfrac{x}{-7}$

17) $21 + x = -1$

18) $-4 = -4 + y$

19) $x - \dfrac{7}{9} = \dfrac{1}{3}$

20) $-7x = 35$

21) $5.1 + x = -1.3$

22) $= \dfrac{x}{0.8} = -1.2$

23) $2.8 + x = 11$

24) $-0.5x = 5$

24) Math Language: Translating Math Expressions

Addition Words: add, plus, sum, increased, all, total, altogether….	**Subtraction Words:** subtract, minus, difference, decreased by, reduced, less…
Multiplication: product, times, twice, multiplied by etc.	**Division:** divide, average, shared evenly or equally, split equally, quotient…

Translate the following math phrases into numerical expressions:

Example: eleven more than twenty $11 + 20$

1. the difference between thirty and fifteen

2. the sum of five, four and three

3. the product of twenty and two

4. fifteen less than 70

5. the quotient of 15 and 7

6. seventeen increased by five

7. eight reduced by seven

8. the quotient of twenty and 5 decreased by four

Translate into verbal expressions:

Example: $11 \times 5 + 20$; the product of eleven and five increased by twenty

9. $5 - 3$

10. $7 + 3$

11. $5/9$

12. $5 \times 3 \div 7$

13. $33 \times 2 + 7$

14. $5 \div 2 - 3$

25) Math Language: Variables

Variables: In math letters such as **x, y, z** etc. are used to represent the unknown variables.

      Example: a number increased by 5 is represented as **x + 5**.

Translate each phrase into an algebraic expression. Use letter "x" as the unknown variable. Use also y for the second variable if there is any.

    1. eleven more than a number

    2. the difference between a number and fifteen

    3. twenty pounds more than his weight

    4. the product of five and a number

    5. fifteen less than the product of five and number

    7. the quotient of 7 and a number

    8. seventeen increased by ten times a number

    9. nine more than the product of three and two

    10. the quotient of twenty and 5 decreased by four.

    11. twice of her age increased by twice of her age

    12. eight times the product of length and width

26) Equations: Word Problems

Example: the sum of 5 and a number is 25. What is the number?
   1) **Build the equation**: 5 + x = 25
   2) **Solve:** since 5 + 20 = 25 ; x must equal 20

Write the equation and then solve.

1) The sum of a number and 3 is ten.

2) Four times a number increased by 2 is 22.

3) The combined age of Ali and Asha is 27. Ali is 14. How old is Asha?

4) A number increased by 35 is 62.

5) A number decreased by 10 is 42.

6) Five times a number is 45.

7) The quotient of 12 and a number 4.

8) Since last year, a baby grows 13 inches to become 30 inches. How tall was the baby?

9) The product of a number and 7 is 63.

10) Omar was 150 pounds a week ago, then he lost 3 pounds before gaining more weight to become 180 pounds. How much he gained?

27) Two Step Equations: Variables on One Side

1) Subtract 15 from both sides	$\begin{array}{r} 2x + 15 = -5 \\ -15 \quad -15 \\ \hline 2x \quad = -20 \end{array}$
2) Divide both Sides by the coefficient of x which is 2	$\dfrac{2x}{2} = \dfrac{-20}{2}$ or $x = -10$
Check your Solution Correct!	$2(-10) + 15 = -5$ $-5 = -5$

1) $2x + 7 = 21$

2) $3x + 2 = -10$

3) $2y - 6 = -2$

4) $9x + 8 = -1$

5) $3x - 3 = -30$

6) $2 = -10x + 12$

7) $-10 = -7x + 4$

8) $21 = 8y + 3$

9) $2x + 11 - 5 = 8$

10) $15x + 40 = -5$

11) $20 + 8x = 4$

12) $4x + 5 = -55$

13) $5x - 5 = 5$

14) $3y - 13 = 2$

15) $-2 = 3x + 1$

	$\dfrac{x}{4} + 8 = 3$
1) Subtract 8 from both sides	$-8 = -8$
	$\dfrac{x}{4} = -5$
2) Multiply both Sides by the denominator which is 4 and simplify	$4 \times \dfrac{x}{4} = 4 \times (-5)$ or $x = -20$
Check your Solution	$\dfrac{-20}{4} + 8 = 3$
Correct!	$3 = 3$

1) $\dfrac{x}{2} + 7 = -3$

2) $\dfrac{x}{3} + 2 = -10$

3) $\dfrac{x}{4} - 6 = -2$

4) $\dfrac{x}{3} + 8 = -1$

5) $\dfrac{x}{-6} - 3 = -10$

6) $2 = -\dfrac{x}{4} + 1$

7) $-10 = \dfrac{-x}{2} + 4$

8) $10 = \dfrac{y}{2} + 3$

9) $\dfrac{x}{3} + 11 - 5 = 8$

10) $\dfrac{x}{5} + 12 = -5$

11) $-10 + \dfrac{x}{-3} = 2$

12) $\dfrac{y}{4} - 5 = -6$

13) $\dfrac{x}{5} - 5 = 5$

14) $\dfrac{y}{5} - 1 = 2$

15) $-2 = \dfrac{x}{2} + 1$

29) The Distributive Property

Follow the examples to simplify the algebraic expressions:

1) $2(3 - 5x) =$
Multiply (distribute) the 2 to both terms: $2 \times 3 + 2(-5x) = 6 - 10x$

2) $-3(2x - 5y) =$
Multiply (distribute) the -3 to both terms : $-3 \times 2x + (-3 \times -5y) = -6x + 15y$

Use distributive property to write the equivalent algebraic expression:
Remember *the* the parenthesis ( ) mean times.

1. $-6(x - 4)$

2. $6(-3 + x)$

3. $-3(z - 4y)$

4. $-4(r + 8s)$

5. $-2(2a - 4b)$

6. $9(-3 + 2x)$

7. $-3(2x - 4y)$

8. $-4(8 - 3y)$

9. $-3(x - y)$

10. $-5(k - 4m)$

11. $3(5 - 4x)$

12. $-4(2x - 8)$

13. $-4(x + y)$

14. $(-2 - 3x)5$

15. $-5(-2y - 3)$

16. $-9(-6y - 5)$

17. $(7 - 4w)(-4)$

18. $-2(-6y - 4x)$

19. $(x - 4)(-3)$

20. $3(-2 - 2w)$

30) Combining Like-Terms

Follow the examples to combine like terms

1) $2x + 15 + 5x = 7x + 15$
2x & **5x** are like terms. So add their coefficients (2+5)x =7x
15 is by itself.

2) $15x - 20x + 5y + 3 - 7$ $= -5x + 5y - 4$
15 & -20 are like terms: $(15 - 20)x = 5x$
3 and -7 are like terms = -4
5y is by itself.

1) $2x + 7x - 21$

2) $8x - 2y - 4x$

3) $y - 5y$

4) $3x - 5x + 2y$

5) $2(-3 + 2x) - 4x$

6) $-3x - 3x + 1$

7) $-2(-10x - 3) + x$

8) $5x - 8x - 2$

9) $-3x + 4x - y$

10) $-2y + 8y + 2x$

11) $-2(y + 2 - y)$

12) $6y - 3 + 9$

13) $2(x - 3) - 4$

14) $20x + 8x - 4x$

15) $-4y - 2y + 10y$

16) $3x - 1 - 4x$

17) $-2(a - 3b) - 4a$

18) $16x^2 + 8x - 4$

19) $-4y + 2y + 10y^2$

20) $-x - 1 - 7x$

31) Add/ Subtract Algebraic Fractions

Follow the example to simplify the algebraic Fractions:

Example	1) $\dfrac{x}{3} - \dfrac{x}{5}$
1. Find the LCM. 2. Make the denominators equal to the LCM by multiplying right numbers. 3. Add/ Subtract and Simplify	*The LCM is 15* $\dfrac{x}{3} - \dfrac{x}{5} = \dfrac{x \times ⑤}{2 \times ⑤} + \dfrac{x \times ③}{5 \times ③}$ $= \dfrac{5x - 3x}{15} = \dfrac{2x}{15}$

1. $\dfrac{5x}{2} - \dfrac{x}{2}$

2. $\dfrac{x}{3} - \dfrac{x}{4}$

3. $\dfrac{2x}{3} + \dfrac{x}{9}$

4. $\dfrac{3x}{5} - \dfrac{2x}{5}$

5. $\dfrac{4y}{7} - \dfrac{2y}{5}$

6. $\dfrac{3w}{4} - \dfrac{w}{5}$

7. $\dfrac{9t}{7} + \dfrac{2t}{14}$

8. $\dfrac{3k}{4} - \dfrac{2k}{3}$

9. $\dfrac{t}{2} + \dfrac{2t}{5}$

10. $\dfrac{2x}{7} - \dfrac{2x}{5}$

11. $\dfrac{m}{3} - \dfrac{7m}{4}$

12. $\dfrac{2x}{4} - \dfrac{5x}{16}$

13. $\dfrac{t}{3} + \dfrac{t}{3}$

14. $\dfrac{5x}{9} - \dfrac{x}{4}$

15. $\dfrac{n}{7} - \dfrac{n}{7}$

16. $\dfrac{4x}{3} - \dfrac{5x}{4}$

32) Solve Equations: Variables on Both Sides I

Follow the example to solve equations:

Example: Solve for x:	$9x = 3x + 12$	$2 + x = 3x + 12$
Step1: Get the variable on one side. Subtract the smaller from both sides. Step 2: Simplify	$9x - (3x) = 3x - (3x) + 12$ $6x = 12$ $x = 2$	$2 + x - (x) = 3x - (x) + 12$ $2 = 2x + 12$ $2 - (12) = 2x + 12 - (12)$ $-10 = 2x$ $-5 = x$

1) $7x = 2x + 10$ 2) $3x = 14 - 4x$ 3) $y - 5y = 12$ 4) $3\,x = 5x + 16$

5) $2(-4 + x) = 4x$ 6) $-x + 4 = 3x + 12$ 7) $-2(5x - 3) = 2x$ 8) $5x = 8x - 3$

9) $3x = 4x + 1$ 10) $-2y = 8y + 2$ 11) $-2(y + 2) = 2y$ 12) $6y = 3 + 9y$

13) $2(x - 3) = 4x$ 14) $20x = 8x - 24$ 15) $4y - 2y = 10y + 8$ 16) $3x - 1 = 4x$

17) $-2(a - 3) = 4a$ 18) $16 + 8x = 4x$ 19) $4y + 2 = 10y + 4$ 20) $-x - 1 = 7x$

33) Solve Equations: Variables on Both Sides II

Follow the example to solve equations:

$8x + 10 = -6x + 12$
$8x + (6x) = -6x + (6x) + 14$
$14x = 14$
$x = 1$

$10 - x + 5 = -8x - 6$
$15 - x + (8x) = -8x + (8x) - 6$
$15 + 7x = -6$
$15 - 15 + 7x = -6 - 15$
$7x = -21$
$x = -3$

1) $4x - 5 = 2x + 10$

2) $7 - 3x = 11 - 2x$

3) $8y - 3y = 2y + 6$

4) $2 - 4(2 + x) = 14x + 12$

5) $-10x + 6 = 3x + 15$

6) $-16 + 4x = 2x + 8$

7) $-15 - 6x = -4x + 3x$

8) $9 - 4y = 7y - 13$

9) $-2(-x - 8) = -2x$

10) $2 - 5(x - 3) = -4x$

11) $20x - 6x = -4x - 24$

12) $10 - 4y - 2y = 4y$

13) $-20(a - 3) = -4a - 12$

14) $16 - 8x = -4x - 4$

34)   Solving Proportions Equations

1) ⇨ A proportion represents two ratios (fractions) that are equal:

2) **To solve a proportion:** multiply the diagonal numbers or variables & solve:

$\dfrac{2+x}{5} = \dfrac{6}{2}$	$\dfrac{y}{5} = \dfrac{9}{7}$	$\dfrac{9}{2} = \dfrac{3p}{6}$
$2(2+x) = 30$ $4 + 2x = 30$ $x = 13$	$7y = 5 \times 9$ $y = 45 \div 7$ $y = \dfrac{45}{7}$	$6p = 54$ $p = 54 \div 6$ $p = 9$

1) $\dfrac{x}{4} = \dfrac{7}{2}$

2) $\dfrac{3x}{8} = \dfrac{7}{2}$

3) $\dfrac{x+4}{5} = \dfrac{6}{2}$

4) $\dfrac{2x+2}{6} = \dfrac{6}{7}$

5) $\dfrac{4}{x+3} = \dfrac{6}{5}$

6) $\dfrac{4}{4+y} = \dfrac{2}{3}$

7) $\dfrac{9}{4} = \dfrac{y+7}{2}$

8) $\dfrac{5}{4} = \dfrac{y+3}{2}$

9) $\dfrac{9}{5} = \dfrac{y+5}{2}$

10) $\dfrac{3}{4} = \dfrac{12}{y+3}$

11) $\dfrac{1}{5} = \dfrac{21}{5y}$

12) $\dfrac{1}{4} = \dfrac{7}{2y-4}$

13) $\dfrac{t}{0.4} = \dfrac{1+t}{0.8}$

14) $\dfrac{4}{5} = \dfrac{9+n}{n-6}$

15) $\dfrac{8x-2}{0.6} = \dfrac{8}{0.4}$

35) How to Solve Proportion Word Problems:

Example 1: Suban runs each morning 4 miles in 15 minutes. How many minutes will it take her to run 18 miles?

Step1: Use variable, like x, d etc. for the missing number and set up the proportion (4 miles in 15 Minutes means divide).	$\dfrac{4 \text{ miles}}{15 \text{ minutes}} = \dfrac{18 \text{ miles}}{x \text{ minutes}}$
Step 2: **Cross Multiply**:	$15 \times 18 = 4\,x$
Step3: **Solve the proportion** (Isolate variable *x*)	$\dfrac{15 \times 18}{4} = \dfrac{4x}{4}$ $x = 67.5 \; minutes$

We can exchange denominators and numerators as far as the corresponding units are in the same side). Try this and call me if you do not get the same answer as before!)

$$\frac{15 \text{minutes}}{4 \text{ miles}} = \frac{x \text{ minutes}}{18 \text{ miles}}$$

Example 2:

A scale drawing 2 inches represent 30 miles. What distance does a line segment of 7 inches represent?

Step1: Set up the proportion (Let's call d the distance)	$\dfrac{2 \text{ inches}}{30 \text{ miles}} = \dfrac{7 \text{ inches}}{d}$
Step 2: Cross Multiply:	$2d = 7 \times 30$
Step3: Solve the proportion (Isolate the d)	$\dfrac{2d}{2} = \dfrac{210}{2}$ $d = 105 \; miles$

Solve the following Proportion Problems:

1) There was just 2 computers shared by every 5 students. If there were 30 students in class, how many computers they share?

2) Seven students out of every 10 Pre-university students who took the math test passed the test. This month 80 will take the test. How many are expected to pass the test.

3) It takes 2 cubs of sugar to prepare 50 cakes. How many cakes that can be prepared from five and half cubs of Sugar?

4) On a typical school day 5 students in every 75 students claim to have lost their pencils. How many students that will claim to have lost their pencils in a school of 240 students?

5) Gloria runs 6 miles in 30 minutes. At that rate, how far could she run in 120 minutes?

6) Assume test grades are proportional to study time. Asma has studied 4 hours per week to score 80. How many hours should Asma study math per week if she wants to get 98?

7) Waris is preparing cake. She knows that every six cups of flour needs one cup of sugar. If she wants to use 36 cups of flour, how many cups of sugar she should have used?

8) Jack drove his Honda for 200 miles and used 5 gallons of fuel. How long he traveled if he used 18 gallons?

36) Solve and Graph Inequality

1) Solve Inequality just like you do in the equality but use the inequality sign:

Less <,   Less or equal ≤,   Greater >, Greater or Equal ≥

2) See these examples:

1) $x + 3 \geq 1$	$x + 3 + (-3) \geq 1 + (-3)$ *Or* $x \geq -2$ See the interval is closed (dark)	
2) $-4 < x \leq 2$	See how one side is open and the other side is closed	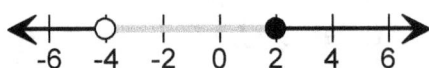
3) $-2x \leq -8$	$\dfrac{-2x}{-2} \geq \dfrac{-8}{-2}$ *or* $x \geq 4$ **Note** how the inequality **changes direction** when multiplied or divided by negatives.	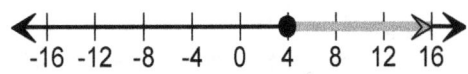

Solve (if needed), then graph the inequality

1) $x \geq 1$

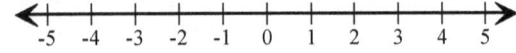

2) $y - 5y \leq 12$

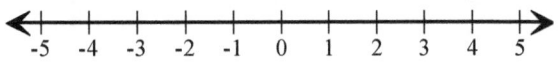

3) $-3 < x \leq 1$

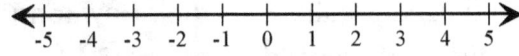

4) $5x < 8x - 3$

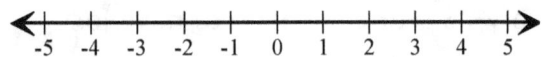

5) $3x + 6 > 2x + 1$

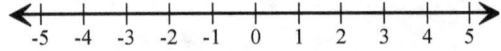

6) $-2 \leq x \leq 1$

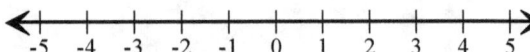

37) Write the corresponding Inequality

1)

$x > 2$

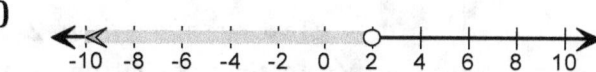

2)

3)

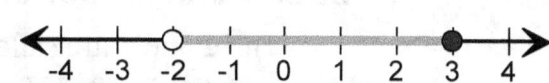

4)

5)

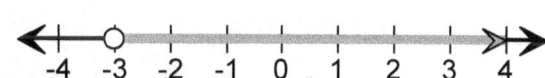

6)

7)

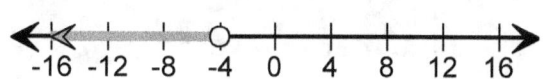

8)

9)

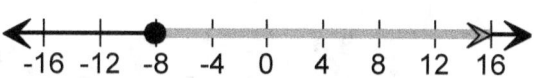

10)

Solve and draw the inequality

11) $x + 5 \geq 1$

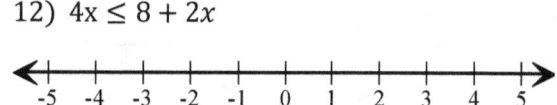

12) $4x \leq 8 + 2x$

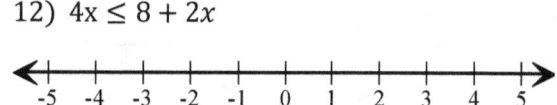

13) $-2w - 4 \leq 4$

14) $3x - 12 < 3$

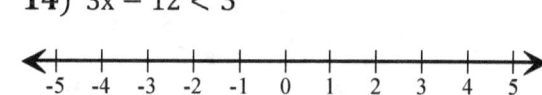

15) $-3x - 6x > -18$

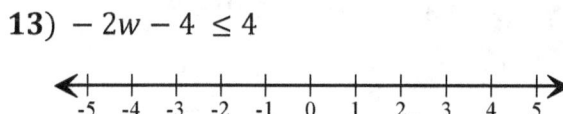

16) $-2y + 6 < -2$

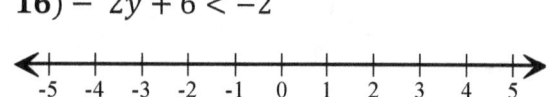

38) Exponents: The Basics

1. **An exponent represents repeated multiplications:**
   - ✓ Instead of writing: $2 \times 2 \times 2 \times 2$, Simply write: 2^4
   - ✓ Instead of writing: $y.y.y.y.y$, Simply write y^5

2. **The base and the power:**
   - ✓ The **2** and y are called bases
   - ✓ The **4 & 5** are the exponents

3. **Read it :**
   - ✓ 5^3 as, five to the third power
   - ✓ y^4 as, y to the fourth power

4. **Zero power:**
   - ✓ *Any number that has **zero** as an exponent is 1*
   - ✓ *Therefore,* $(-127)^0 = 1$; $y^0 = 1$; *any number$^0 = 1$*

5. **To add or subtract variables:**
   - ✓ Add only same variables with the same power. $3x^3 + 2x^3 = 5x^3$
   - ✓ Never add a variable and a number: $2x^2 + 5 = 2x^2 + 5$

6. **Never add two powers:**

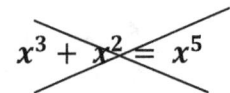

$$x^3 + x^2 = x^5$$

Exercise: *Simplify*

1) $9x^2 + 4x^2 =$

2) $5y^2 - 3y^2 + y =$

3) $2x^3 + x^0 + 5 =$

4) $3^3 + 2x^3 + 5x^3 =$

5) $10y^2 - 3y^2 + 2x^2 =$

6) $7^3 - 5x^2 + x =$

39) Add or Subtract Exponents

Example 1: $3^4 + x^2$ $= 3 \times 3 \times 3 \times 3 + x^2$ $= 81 + x^2$	Example 2: $5^2 + 7^2$ $= 5 \times 5 + 7 \times 7$ $= 25 + 49$ $= 74$	Example 3: $3x^2 + 6x^2 =$ $(3 + 6)x^2 = 9x^2$

(1) $10^0 =$

(2) $3^0 + 1^0 + x^0 =$

(3) $5^3 - 1^0 =$

(4) $3^4 - 3^0 =$

(5) $5^2 - 6 =$

(6) $9x^2 + 7x^2 =$

(7) $5y^7 + 4y^7 =$

(8) $0^5 + 2^2 =$

(9) $3^5 + 2^2 =$

(10) $5y + 3y^2 - 2y =$

(11) $0^5 + 10\,y^4 - 7y^4 =$

(12) $4^4 + 2^2 + 17^0 =$

(13) $10y^4 - 17y^4 =$

(14) $b^5 + 10\,y^4 - b^5 =$

(15) $4y^3 + 2y^2 + 17y^2 =$

40) Multiply and Divide Exponents

Simplify the Exponents. Follow the examples:

$2^3 \times 2^2 \Rightarrow 2^{3+2} = 2^5$	$x^5 (x^3) \Rightarrow x^{5+3} = x^8$
$\dfrac{3^7}{3^2} = \Rightarrow 3^{7-2} = 3^5$	$\dfrac{y^8}{y^3} = \Rightarrow y^{8-3} = y^5$

(1) $3^4 \times 3^4 =$

(2) $x^4 \times x^8 =$

(3) $\dfrac{7^3}{7} =$

(4) $10^0 \times 3^4 =$

(5) $12^0 \times 12^3 =$

(6) $\dfrac{3^7}{3^2} =$

(7) $y^0 \times y^4 =$

(8) $\dfrac{m^7}{m^3} =$

(9) $5^2 - \dfrac{10^2}{10} =$

(10) $5 \times 3^2 - 27^0 =$

(11) $\dfrac{3^7}{3^2} - 3^5 =$

(12) $\dfrac{x^7}{4^3} \times \dfrac{x^7}{4^2} =$

41) Power of Power (Exponent of an Exponent)

Simplify the Exponents. Follow the examples:

$(x^3)^2 = \Rightarrow x^{3\times2} = x^6$	$x^5(x^3)^2 \Rightarrow x^5(x^6) = x^{11}$
$(y^4)^2 = \Rightarrow y^{4\times2} = y^8$	$\dfrac{y^{12}}{(y^5)^2} = \Rightarrow y^{12-10} = y^2$

(1) $(x^3)^3 =$

(2) $(x^5)^2 =$

(3) $(y^3)^2 =$

(4) $(x^5)^0 =$

(5) $(y^0)^2 =$

(6) $(2^3)^2 =$

(7) $\dfrac{y^6}{(y^2)^2} =$

(8) $3^2(2^2)^2 =$

(9) $x^5(x^5)^2 =$

(10) $2^0(2^4)^2 =$

(11) $(3^2)^2 =$

(12) $\dfrac{y^{12}}{(y^5)^0} =$

42) Negative Exponents

Simplify the Exponents and change all **negative exponents** into positive.
Follow the examples:

$$y^{-4} = \frac{1}{y^4}$$

$$2^{-3} = \frac{1}{2^3} = \frac{1}{8}$$

$$y^{-4} \times y^7 = y^{-4+7} = y^3$$

$$\frac{x^{-5}}{x^2} = \frac{1}{x^2} \times \frac{1}{x^5} = \frac{1}{x^7}$$

(1) $x^{-3} =$

(2) $(x^{-5})^2 =$

(3) $(y^{-3})^2 =$

(4) $\frac{1}{y^{-9}} =$

(5) $(y^{-4})^0 =$

(6) $(2^{-3})^{-2} =$

(7) $\frac{x^{-5}}{x^{-2}} =$

(8) $\frac{x^{-5}}{x^{-2}} =$

(9) $\frac{x^{-7}}{x^{-2}} =$

(10) $\frac{x^9}{x^{-5}} =$

(11) $\frac{x^{-5}}{x^{-5}} =$

(12) $\frac{x^{-5}}{x^{-6}} =$

43) Evaluate Exponents

Evaluate the following exponents. Follow the examples:

⇨	$x^3 \times 3^2$ if $x = 2$
	$= (2)^3 \times 3^2$
	$= 2 \times 2 \times 2 + 3 \times 3$
	$= 8 + 9$
	$= 17$

⇨	$\dfrac{y^7}{y^4}$ when $y = 3$
	$y^{7-4} = y^3$
	$= 3^3$
	$= 27$

(1) $x^4 = $ _____
         if $x = 3$

(2) $\dfrac{m}{7} = $ _____
              if $m = 21$

(3) $m^5 \times 2 = $ _____
       if m=2

(4) $\dfrac{x^7}{3^2} = $ _____
     when x=1

(5) $y^0 \times y^4 = $ _____
       when y =4

(6) $5^2 \times \dfrac{10^2}{10} = $ _____

(7) $n + n^2 - 8 = $ _____
       When n = 4

(8) $\dfrac{x^6}{x^5} \times \dfrac{x^3}{x^2} = $ _____
     When x = 4

(9) $\dfrac{m^5}{m^2} - 81^0 = $ _____
       If m=9

10) $\dfrac{n^4}{n^2} + n^2 = $ _____
       If n=5

44) Scientific Notation-1

Express each number in standard form. See examples below:

	Examples
Since the exponent (4) is positive move the decimal point 4 times to the right. Add zero if you need it.	1) $3.52 \times 10^4 = 35200$ 2) $2.643 \times 10^4 = 26430$
Since the exponents is negative move the decimal point 3 times to the left and 2 times to the left for the second example.	3) $1.47 \times 10^{-3} = 0.00147$ 4) $8.31 \times 10^{-2} = 0.0831$

1. 3.52×10^4

2. 2.8×10^2

3. 5.61×10^{-4}

4. 9.63×10^{-3}

5. 8.02×10^3

6. 7.12×10^4

7. 4.6×10^{-1}

8. 3×10^{-5}

9. 6.5×10^5

10) 5.22×10^6

11) 8.213×10^6

12) 1.61×10^{-5}

13) 3.5×10^{-6}

14) 9.13×10^0

15) 4.21×10^0

16) 3.0×10^1

45) Scientific Notation-2

Express each number in scientific notation. See examples below:

	Examples
First, put a decimal next to the first number. Since the decimal has moved three places to the left make the exponent 3. Remember the decimal for **2643** is hidden at the end. So, it is indeed **2643.0**	$3521.8 = 3.5218 \times 10^3$ $2643 = 2.643 \times 10^3$
First, put a decimal next to the first number. Since the decimal has moved four places to the right make the exponent -4.	$0.000147 = 1.47 \times 10^{-4} =$ $0.000831 = 8.31 \times 10^{-4} =$

1. 0.00045

2. 2.8×10^2

3. 41,000

4. 310

5. 0.0021

6. 1000

7. 610,000

8. 45

9. 0.0032

10. 0.000021

11. 210

12. 0.00031

13. 0.0000013

14. 410

15. 81000

16. 0.56

Do you see from both the table and the graph how many minutes exercised when 12 calories were burned? In math, it is also common to use the **ordered pairs** as below:

(1, 2), (2, 4), (3, 6), (4, 8), (5, 10), (5, 10)

x	y
Minutes in Stair Climber	Calories Burned
1	2
2	4
3	6
4	8
5	10
6	12

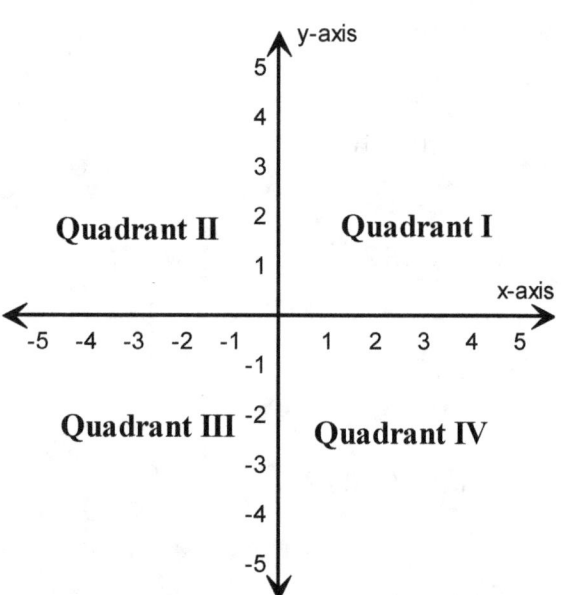

- ✓ The first number of the ordered pair is the **x-coordinate**. The second number of the ordered pair is called the **y-coordinate.**
- ✓ **The Coordinate System**: sometimes also called the Cartesian System is the system that puts ordered pairs in the graph.
- ✓ *The x-axis and y-axis* divide the system into four regions called Quadrants;
- ✓ **The Origin**: is where the x and y-axis meet and is zero: **(0,0)**

47) Graphing: Identify the Ordered Pairs

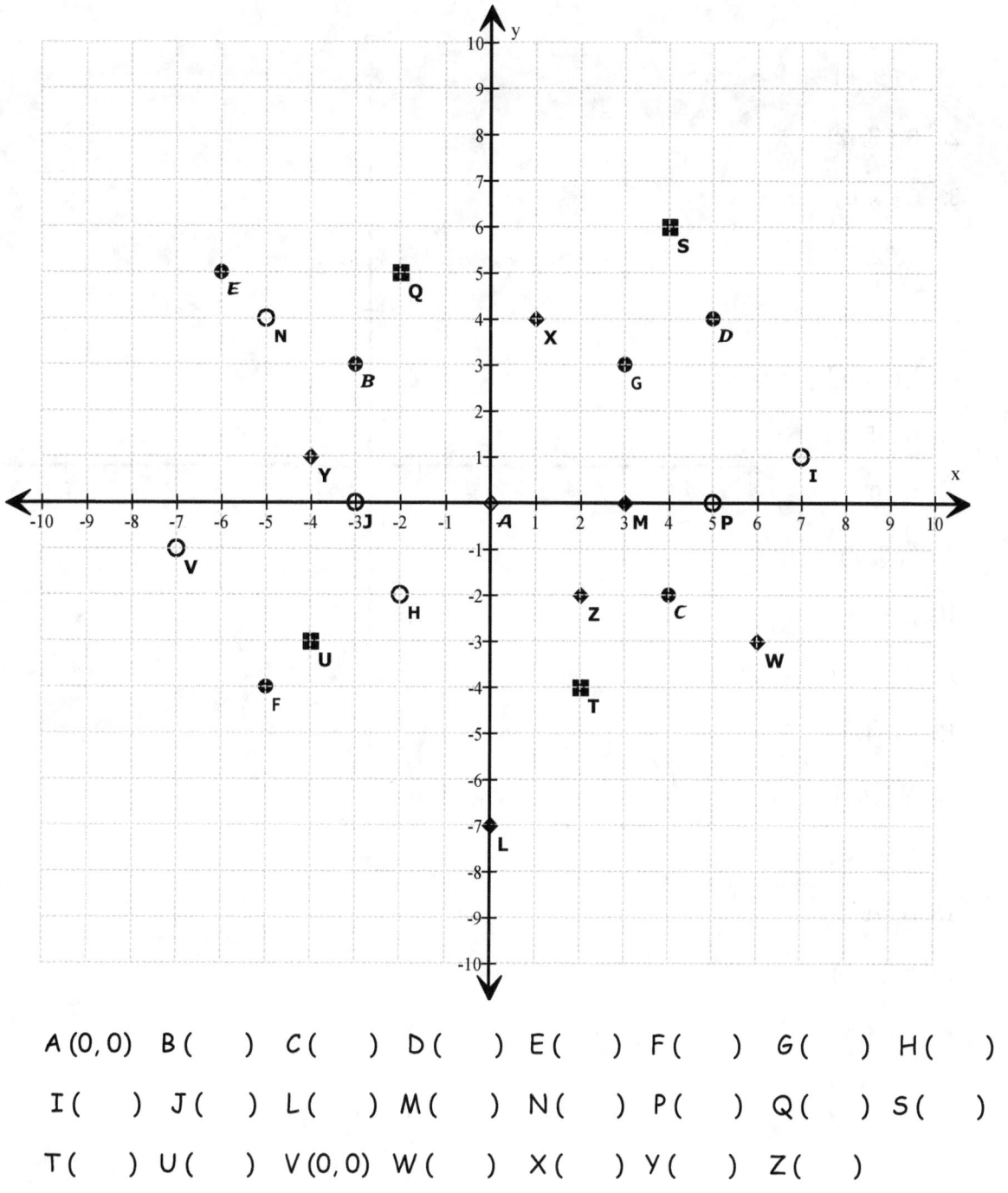

A (0,0) B (   ) C (   ) D (   ) E (   ) F (   ) G (   ) H (   )

I (   ) J (   ) L (   ) M (   ) N (   ) P (   ) Q (   ) S (   )

T (   ) U (   ) V (0,0) W (   ) X (   ) Y (   ) Z (   )

48) Graph ordered pairs in the Coordinate System

1. A (3, 4)

2. B (3, -2)

3. C (3, 1)

4. D (2, 4)

5. E (-3, 4)

6. F (-2, 3)

7. G (-5, 1)

8. H (0, 4)

9. I (3, 0)

10. J (-3, -4)

11. K (-1, -2)

12. L (-4, 4)

13. M (-6, 4)

14. N (-5, 3)

15. O (-5, 4)

16. P (3, 4)

17. Q (3, 4)

18. R (-3, -4)

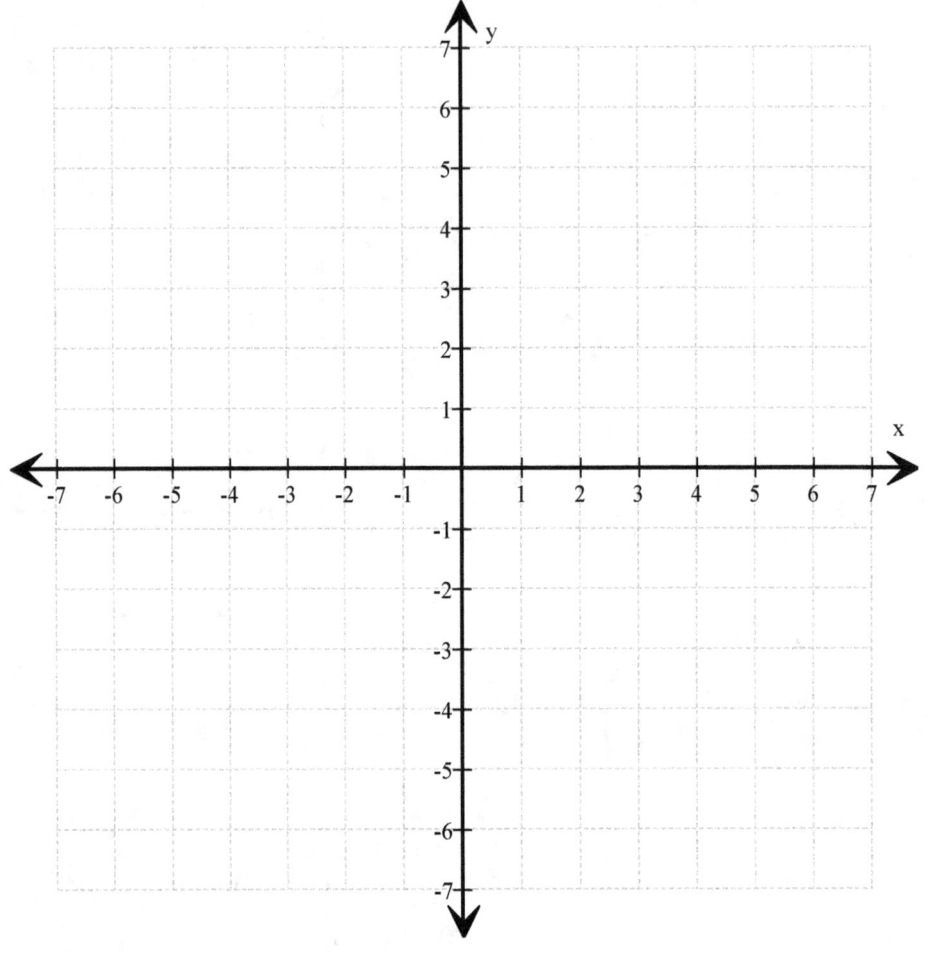

49) PRACTICE: Which coordinate is each person?

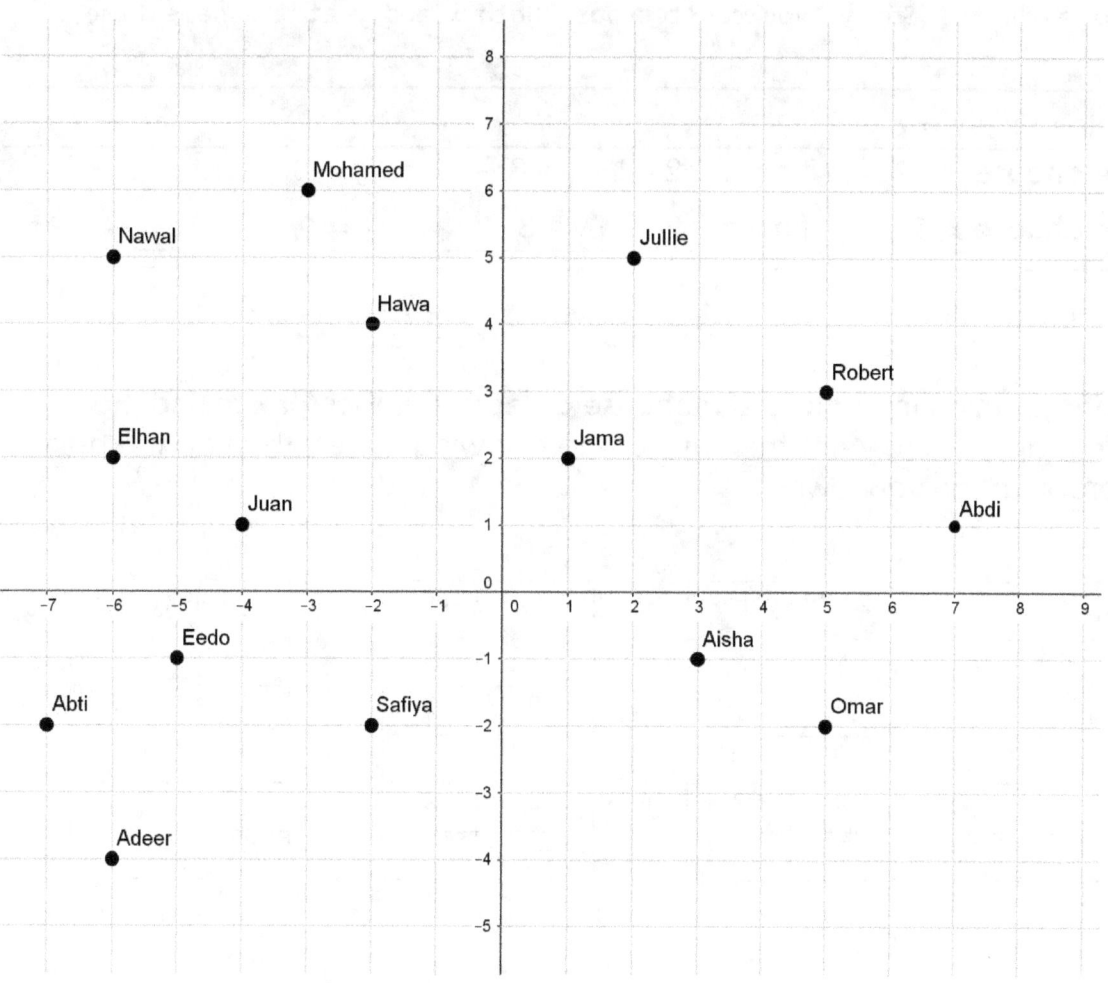

Abti (-7, -2)	Hawa( )	Juan ( )	Omar ( )
Abdi ( )	Eedo ( )	Jullie ( )	Robert ( )
Adeer ( )	Elhan( )	Mohamed ( )	Safiya ( )
Aisha ( )	Jama( )	Nawal ( )	

50) Graphing Linear Equations: How to Find Ordered Pairs

To find the ordered pairs of equation, give any value you like to x, and then solve the y. **(Sorry, you can't choose both x and y at the same time).**

Example : $2x + y = 8$		
If we choose x=0	it means $2(0) + y = 8$	$y = 8$
If we choose x=1	it means $2(1) + y = 8$	$y = 6$

Hint: Choose any number for x but choose an easy number for x to find the y number. Make sure to verify the entry of the following tables. Then do the next page problems on your own!

1) $2x + y = 8$

x	y
0	8
1	6
2	4
3	2

2) $3x - y = 4$

x	y
0	-4
1	-1
2	2
3	5

3) $x - y = -2$

x	y
0	2
1	3
2	4
-1	1

4) $x + 2y = 5$

x	y
0	2.5
1	2
3	1
5	0

5) $y = x - 3$

x	y
0	-3
1	-2
4	1
5	2

6) $y = 3x - 2$

x	y
0	-2
1	1
2	4
-1	-5

7) $y = x + 2$

x	y
0	2
1	3
2	4
3	5

8) $y = -4x - 3$

x	y
0	-3
-1	1
-2	5
1	-7

51) Graphing Linear Equations: Find Ordered Pairs

1) x + y = 1

x	y

2) x - 2y = 4

x	y

3) x - y = 3

x	y

4) 3x + y = -2

x	y

5) y = 2x - 8

x	y

6) y = x - 4

x	y

7) y = x + 1

x	y

8) y = x

x	y

9) y = x - 3

x	y

10) 2x - y = - 2

x	y

11) y = -2x + 4

x	y

12) x + 5y = -10

x	y

13) 3x + y = 6

x	y

14) x - 3y = -3

x	y

15) 2x + y = -8

x	y

16) 2x + y = 0

x	y

52) Find Ordered Pairs for Equations with Fractions

Example 1	$y = \dfrac{1}{2}x + 2$	x	0	2	4	-2	-4
		y	2	3	4	1	0

See! It is easier to choose multiples of the 2 in the denominator for the x such as 2,4, 6, -2, -4...

Example 2	$y = \dfrac{2}{3}x - 4$	x	0	3	6	9	-3
		y	-4	-2	0	2	-6

See again. It is easier to choose multiples of the 2 in the denominator for the x such as 3,6, 9, -3...

1) $y = \dfrac{2}{3}x + 1$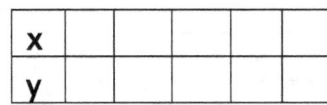

2) $y = \dfrac{2}{5}x - 3$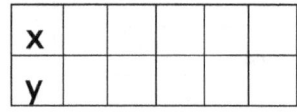

3) $y = \dfrac{1}{4}x + 1$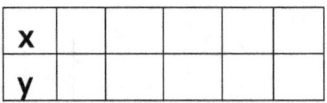

4) $y = \dfrac{3}{2}x - 2$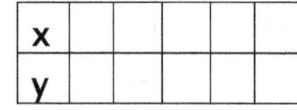

5) $y = \dfrac{1}{3}x + 1$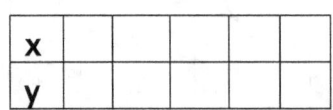

6) $y = \dfrac{1}{5}x - 2$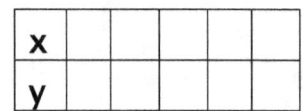

7) $y = \dfrac{3}{4}x + 1$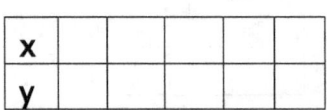

8) $y = \dfrac{4}{3}x - 2$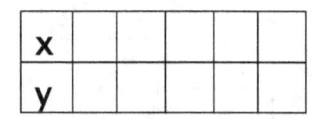

9) $y = \dfrac{1}{6}x - 5$

10) $y = \dfrac{2}{7}x + 1$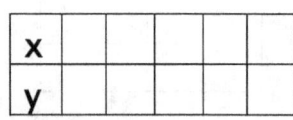

53) Graphing Linear Equation

Prepare 2-4 ordered pairs, 2) **Plot** the point 3) **Graph** the line.

Example 1: **Graph y = 2x+3**

Let's find few ordered pairs
and plot on the graph:

x	y
0	3
-1	1
1	5

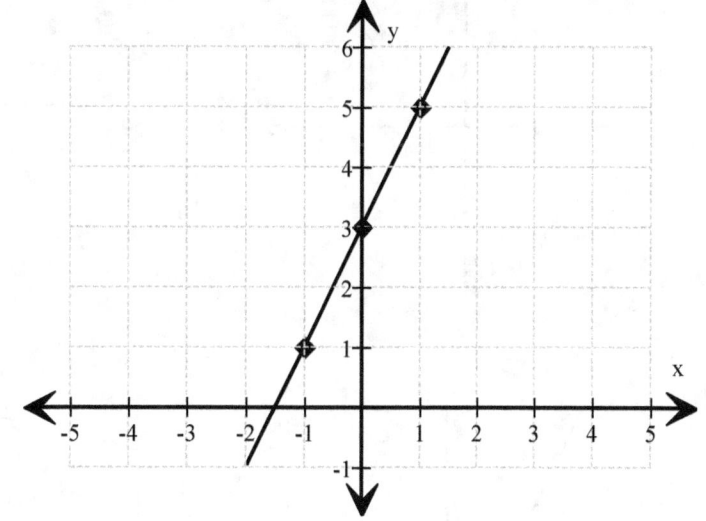

Example 2: **Graph y = -2x+2**

x	Y
0	2
-1	4
1	0

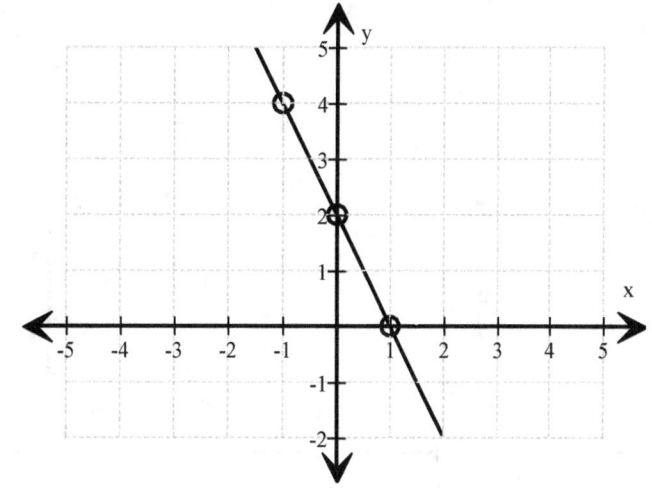

54) Match the Graphs with their correspondent equations

| A) y = 2x-3 | B) y =-2x - 2 | C) y = 3x+1 | D) y = x | E) y =-x+3 | F) y= -4x+3 |

1)

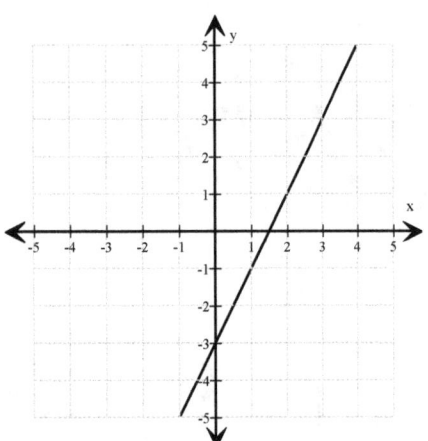

2)

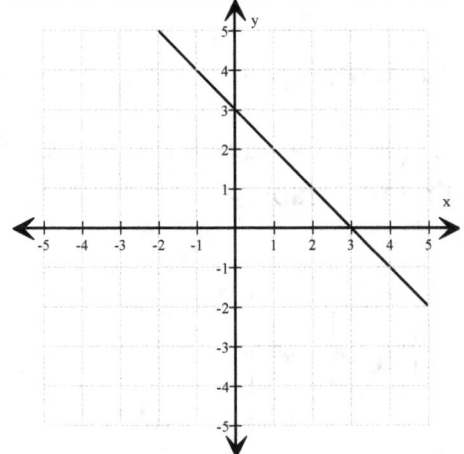

3)

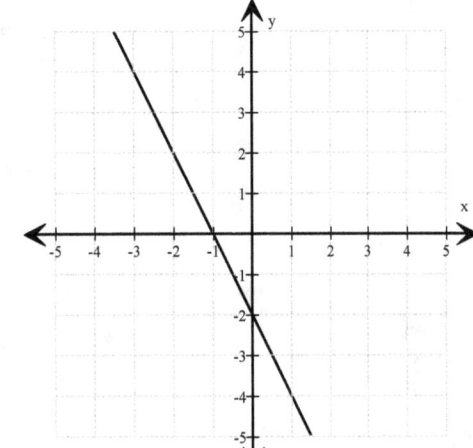

4)

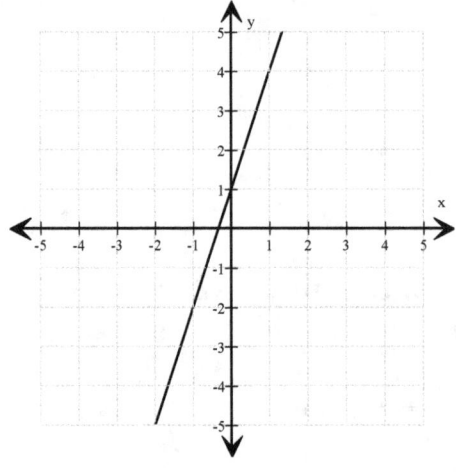

5)

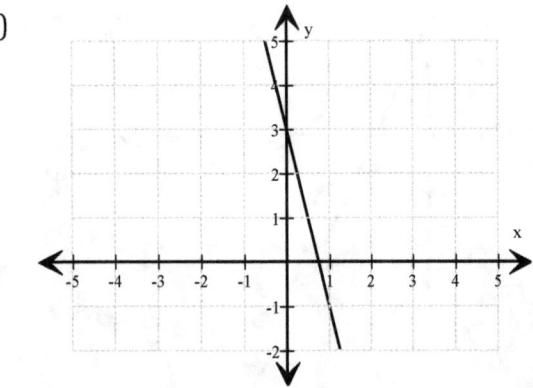

6)

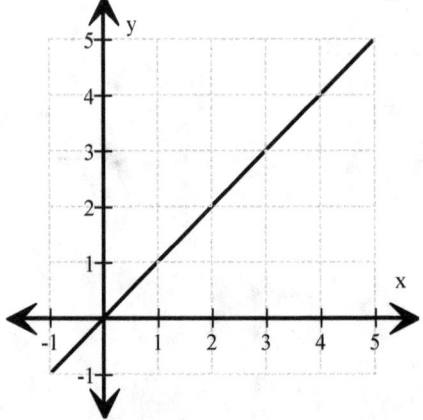

55) Graph These Linear Equations

1. **Prepare** 3 ordered pairs, 2) **Plot** the points 3) **Graph** the line:

1) $y = 2x$

2) $y = x - 2$

3) $y = -4x - 2$

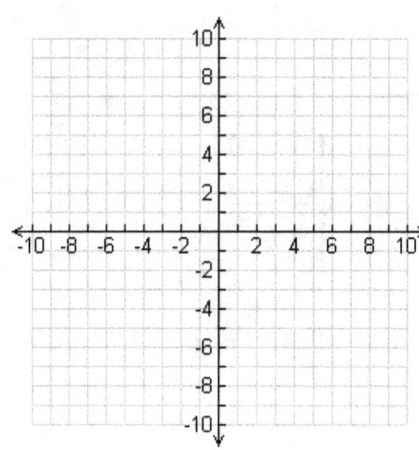 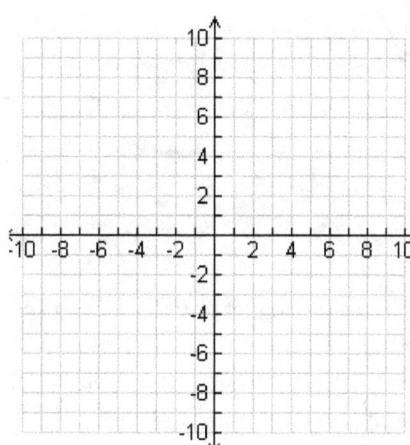 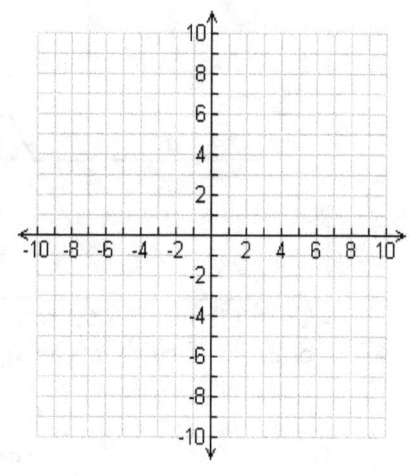

4) $y = x + 2$

5) $y = 3x + 1$

6) $2x + 3y = 6$

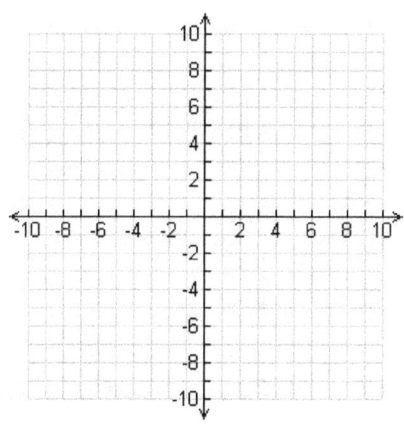 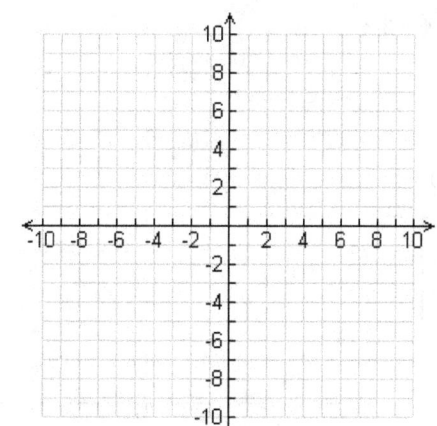 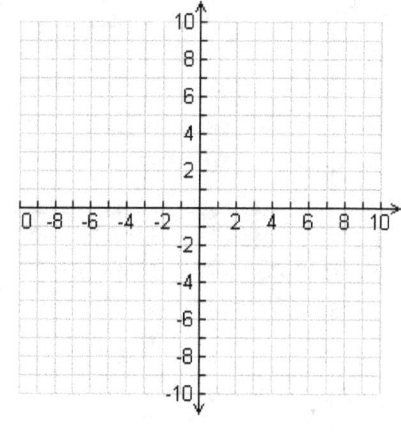

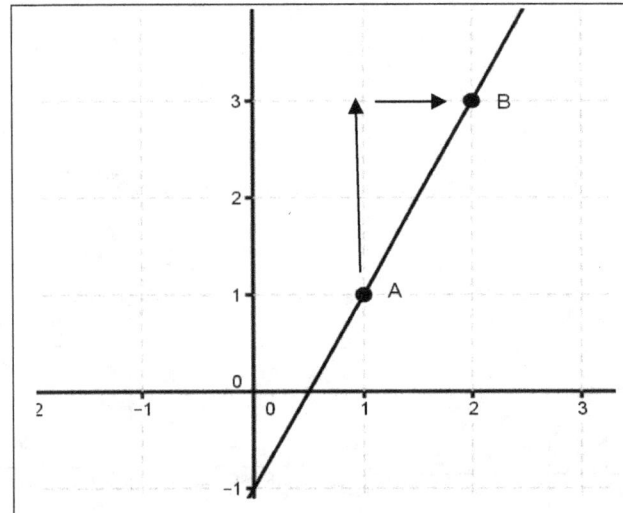

To get from point A to B rise up (at y axis) by 2 points and run (on x axis) by 1:

$$Slope = \frac{Rise}{Run} = \frac{2}{1} = 2$$

Rise from point C and A. = 3

Run from B to C= -1 (It is negative to move backwards)

$$Slope = \frac{Rise}{Run} = \frac{3}{-1} = -3$$

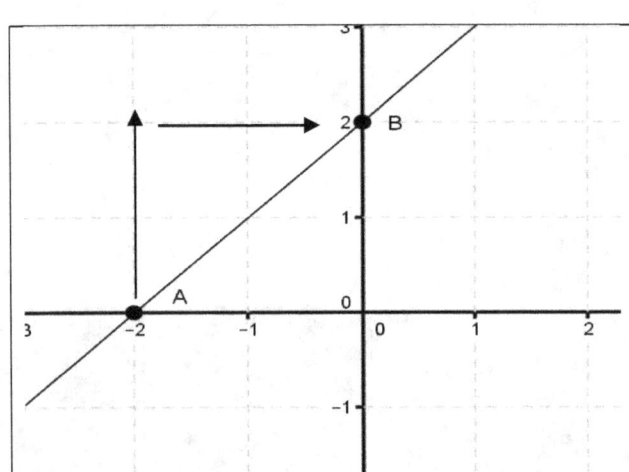

Rise from point A to B. = 2

Run for A to B= 2

$$Slope = \frac{Rise}{Run} = \frac{2}{2} = 1$$

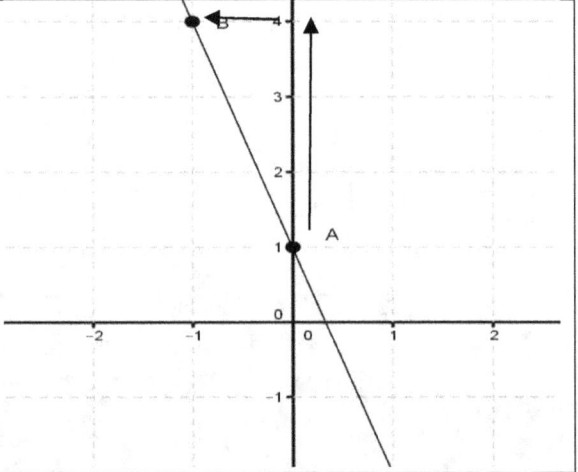

$$\frac{Rise}{Run} = \frac{3}{-1} = -3$$

57) Find the Slope from the Graphs

1)

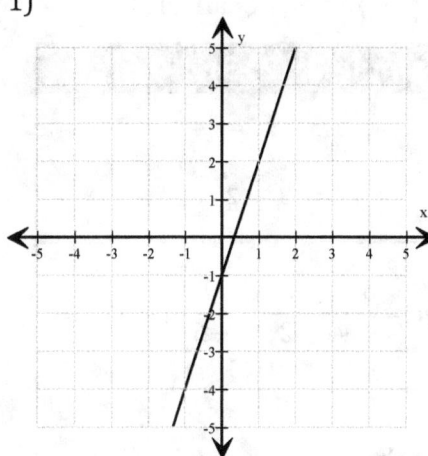

2)

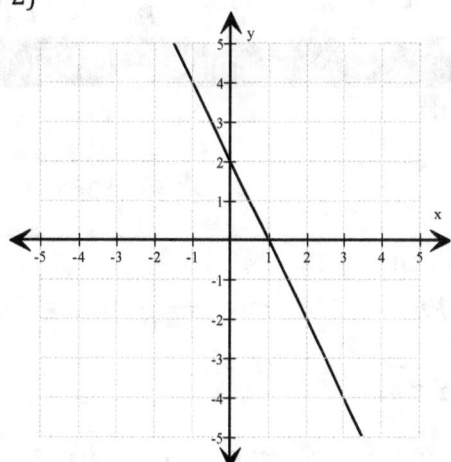

3)

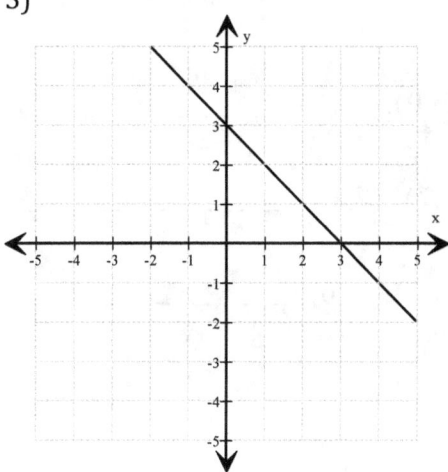

4)

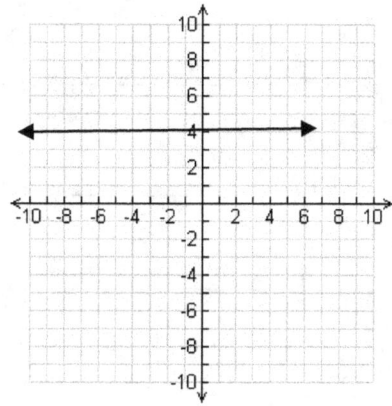

5)

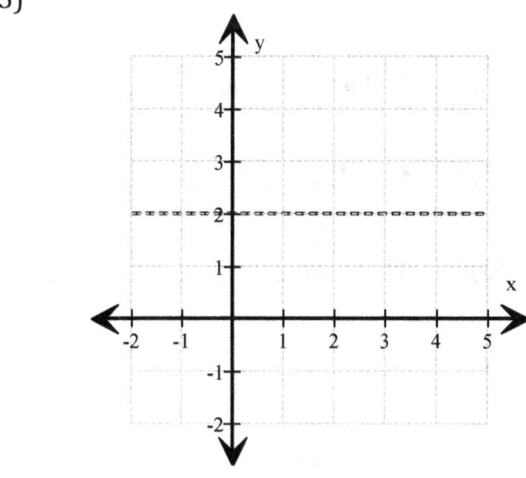

6)

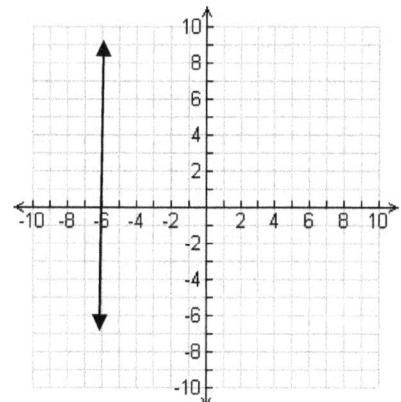

58) Find the Slope from Ordered Pairs

Find the slope that passes through each pair of ordered points: See examples.

	Example1 : (0,3),(4,1)	Example 2: (2,-3), (-4,6)
1) Label it:	$x_1=0$, $x_2=4$ $y_1=3$, $y_2=1$	$x_1=2$, $x_2=-4$ $y_1=-3$, $y_2=6$
2) Put it in the Formula: $m = \dfrac{y_2 - y_1}{x_2 - x_1}$	$m = \dfrac{1-3}{4-0} = \dfrac{-2}{4} = -\dfrac{1}{2}$	$m = \dfrac{6-(-3)}{-4-2} = \dfrac{9}{-6} = -\dfrac{3}{2}$

Remember: m stands for slope. Also, watch out for the signs!

1. (2, 3), (4,1) _____

2. (-4, 3), (5,1) _____

3. (1, 3), (2,4) _____

4. (5, 5), (6, 7) _____

5. (3, 3), (-4, 1) _____

6. (2,-3), (4, 7) _____

7. (3,-3), (-4, 6) _____

8. (7,-3), (-4, 8) _____

9. (1, 4), (4, 7) _____

10. (0,-3), (-4, 5) _____

11) (2,-0), (-1, 2) _____

12) (2,7) , (7, 6) _____

13) (7,-3), (6, 6) _____

14) (7,-3), (0, 6) _____

15) (2,-3), (-4, 6) _____

16) (5,-3), (3, 6) _____

17) (2,-3), (1, 2) _____

18) (4,-3), (3, 8) _____

19) (2,-3), (-4, 9) _____

20) (2, 9), (-5, 6) _____

59) Write the Equation of the Line: Use Slope/Y-intercept:

A) Use the given slope and the y-intercept to **write the equation of the line** in Slope Intercept form $(y = mx + b)$. See examples:

Example1 :
Slope =3   y-intercept =1
$y = mx + b$
$y = 3x + 1$

Example 2 :
Slope =-1 y-intercept =0
$y = mx + b$
$y = -x + 0$
Or simply: $y = -x$

Remember: "m" is the slope and "b" stands for the y-intercept.

1) Slope =2, y-intercept = 1

2) Slope =-2, y-intercept = 2

3) Slope =-4, y-Intercept = -2

4) Slope =-1, y-intercept = 7

5) Slope = 5, y-Intercept = 0

6) Slope =-6, y-intercept = 11

7) Slope = 0, y-Intercept = -5

8) Slope =$\frac{2}{3}$, y-intercept = -38

9) Slope = -1, y-Intercept = -3

10) Slope =-10 , y-Intercept = -27

B) Given the equation of the line, find the slope and the Y-intercept. See Example:

11) $y = 3x + 1$	$m = 3 ;\ b = 1$	12) $y = 7$	
13) $y = -5x + 7$		14) $y = 8x$	
15) $y = -6x - 4$		16) $y = -3x - 1$	
17) $y = x$		18) $y = 2x + 3$	
19) $y = -9x + 8$		20) $y = -4x - 5$	

60) Graph the Line from slope and y-intercept

Example: Graph the line with the given equation;
$$y = 2x - 1$$
First: see the slope =2 and the y-intercept =-1
So, mark the y-intercept (or -1).

Next: go up 2 points and over 1 point to the right.
1) Connect the point (dots)!

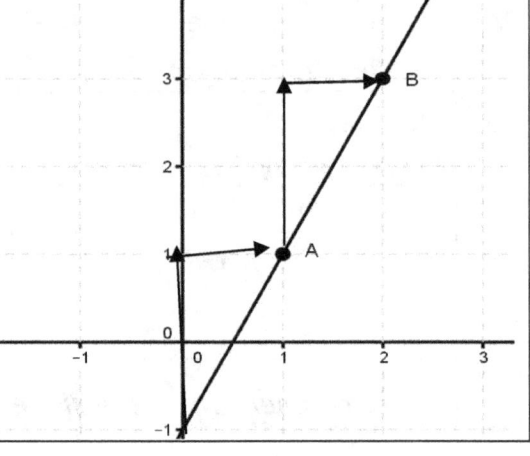

$y = 4x - 2$

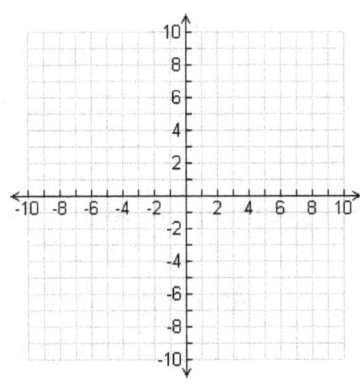

$y = x + 1$

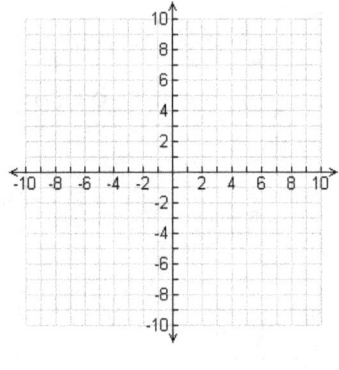

$y = 2x + 2$

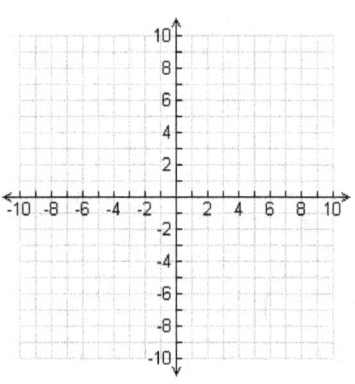

$y = -2x - 1$

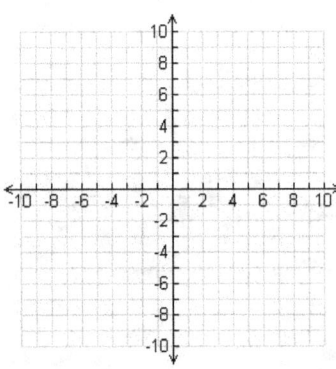

$y = 3x$

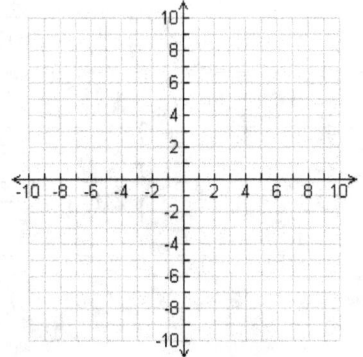

$y = 6x - 6$

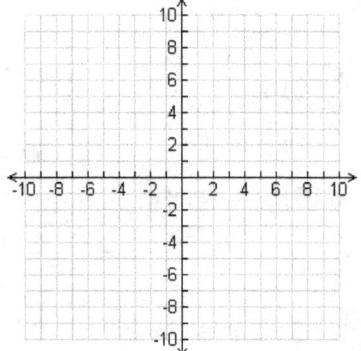

61) Find the slope and the Y-intercept from the Graph

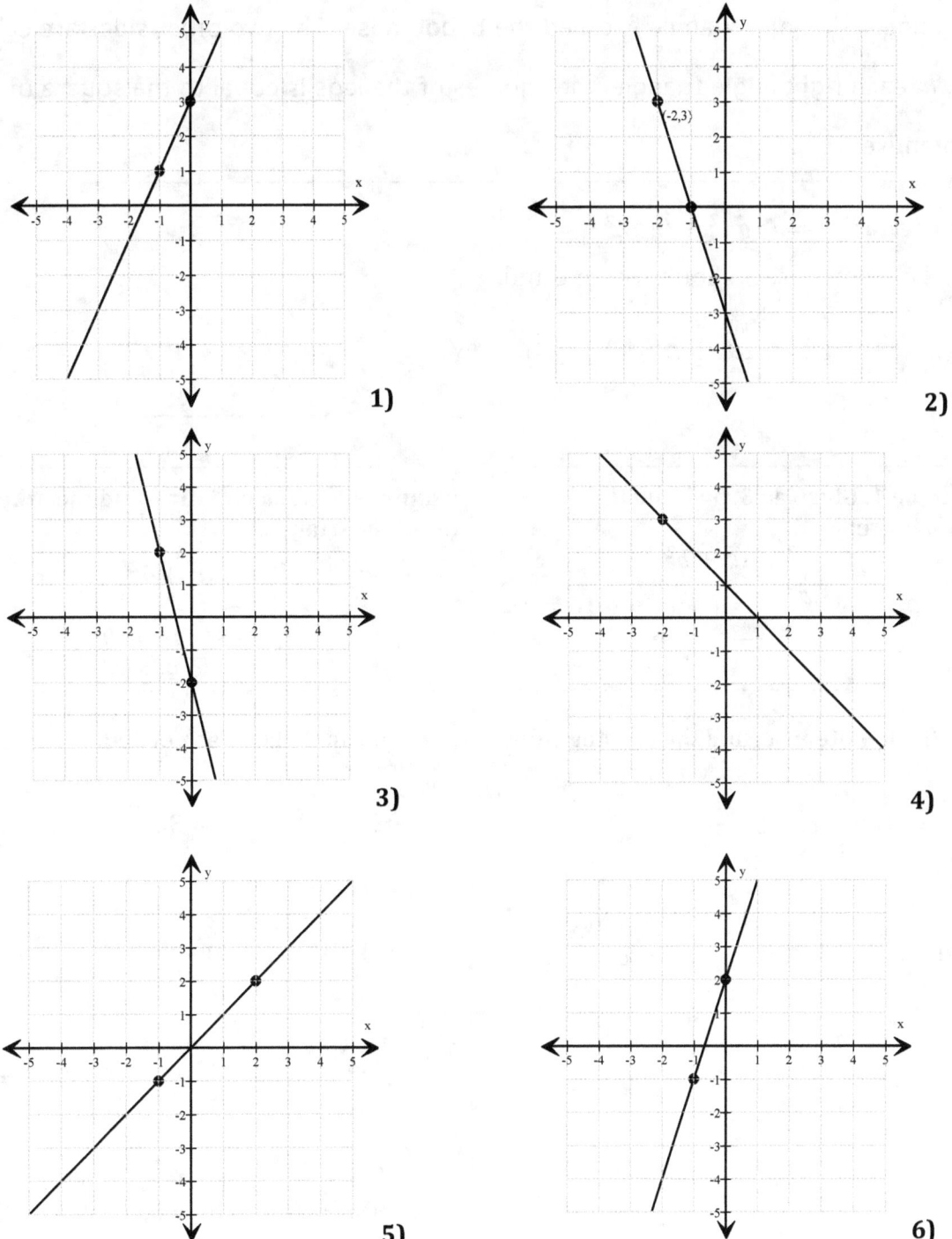

1)

2)

3)

4)

5)

6)

62) The Pythagorean Theorem

The longest side of a triangle is called the **hypotenuse**. The two other sides are called **legs.** In each right angle triangle, the squares of the legs is equal to the square of hypotenuse.

$$hypotenuse^2 = leg1^2 + leg2^2$$

Mostly hypotenuse is shortened for c and the legs as a, and b:

$$c^2 = a^2 + b^2$$

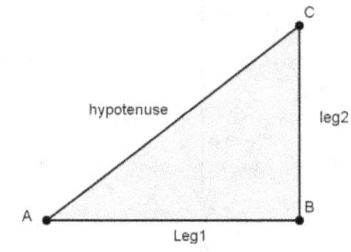

Example 1: Given a= 3, b=4 find the hypotenuse:

$$c^2 = a^2 + b^2$$
$$c^2 = 3^2 + 4^2 = 9 + 16 = 25$$
$$c = \sqrt{25} = 5$$

Example 2: Given c= 5, b=4 find the other leg of the triangle:

$$5^2 = a^2 + 4^2$$
$$5^2 - 4^2 = a^2$$
$$a^2 = 25 - 16 = 9$$
$$a = \sqrt{9} = 3$$

If c is the hypotenuse, find the missing sides. Use two decimal places as needed.

1) $a = 8$ $b = 6$ $c =?$

2) $a =?$ $b = 30$ $c = 40$

3) $a =?$ $b = 6$ $c = 9$

4) $a =?$ $b = 5$ $c = 8$

5) $a = 15$ b? $c = 20$

6) $a = 36$ $b =?$ $c = 49$

7) $a = 7$ $b = 4$ $c =?$

8) $a = 8$ $b = 6,$ $c =?$

63) The Distance Formula

Example: What is the length of segment (a) or the distance between points A and B?

The distance between any two points is calculated by using the distance formula

$$distance = \sqrt{(x_2 - x_1)^2 + (y_2 - y_1)^2}$$

$$d = \sqrt{(5 - 1)^2 + (4 - 1)^2}$$

$$d = \sqrt{16 + 9}$$

$$d = 5$$

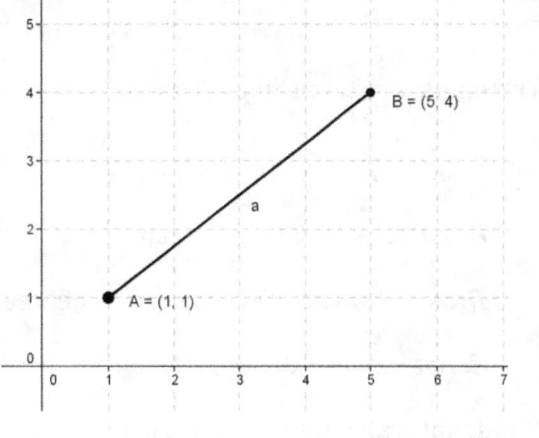

Calculate the distance of the following pairs of points. Use a calculator if you need to!

1. A (1,2), B(3,4)

2. C (4,5), D (6,4)

3. E (7,2), F(4,6)

4) G (-2,-3), H(7,-6)

5)I (-4,5), J (-8,8)

6) K(-3,-3), L(6,6)

7) M (-5, -6) N (2,4)

8) P (-6,4), Q (4,8)

9) R (5,10) S (-5,-6)

64) POLYGONS IN SNAP SHOT!

Definition: A polygon means a closed plane figure with three sides or more. So, we can generally divide it into three groups: triangles (three sided) Quadrilaterals (four Sides), and Figures with 5 sides or more!

1) Triangles: Are three Sides Figures. They could include Equilateral, Isosceles, right and Scalene

> *EQUILATERAL-* All Sides are equal
>
> *ISOSCELES-* 2 sides are equal
>
> *RIGHT ANGLED:* One angle is 90 degrees
>
> *SCALENE-* All sides are Different

2) Quadrilaterals: Are Four sides figures. They Include: (a) **parallelogram** (rectangle, rhombus, square), (b) Trapezoid, and (c) **Kite**

> *PARALLELOGRAMS:* Each opposite sides are equal. Opposite angles are also equal
>
> *TRAPEZOID:* has only one pair of parallel sides
>
> *KITE:* Has 2 pairs of adjacent sides that are equal.

3) Five Sided and More: Examples of common figures of these are

PENTAGON (5 sides)	*HEXAGON* (6 sides),
HEPTAGON (7 sides)	*OCTAGON* (8 sides),
NONAGON (9 sides)	*DECAGON* (10 sides)

Polygons In Three Groups

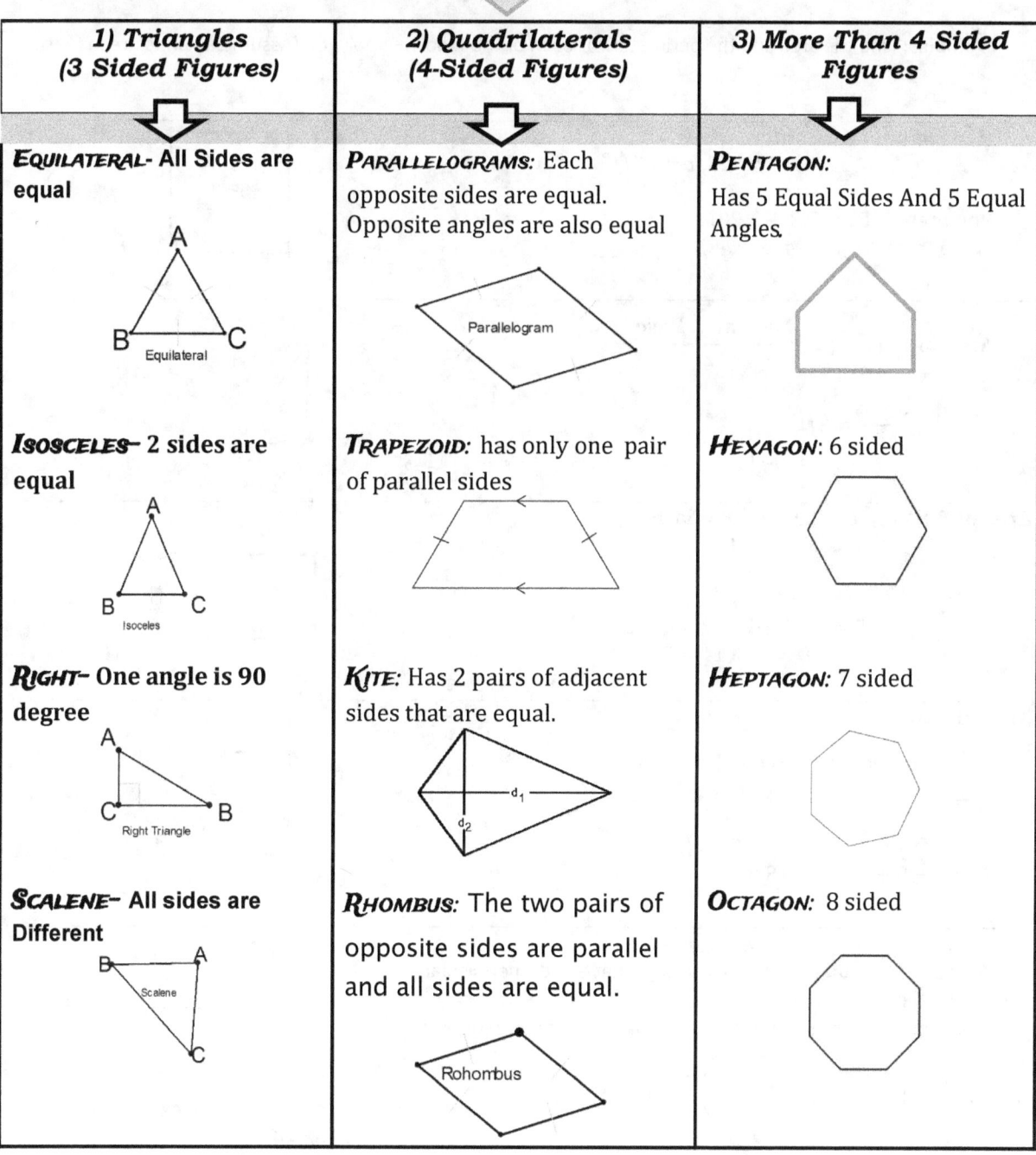

1) Triangles (3 Sided Figures)	2) Quadrilaterals (4-Sided Figures)	3) More Than 4 Sided Figures
EQUILATERAL- All Sides are equal	**PARALLELOGRAMS:** Each opposite sides are equal. Opposite angles are also equal	**PENTAGON:** Has 5 Equal Sides And 5 Equal Angles.
ISOSCELES- 2 sides are equal	**TRAPEZOID:** has only one pair of parallel sides	**HEXAGON:** 6 sided
RIGHT- One angle is 90 degree	**KITE:** Has 2 pairs of adjacent sides that are equal.	**HEPTAGON:** 7 sided
SCALENE- All sides are Different	**RHOMBUS:** The two pairs of opposite sides are parallel and all sides are equal.	**OCTAGON:** 8 sided

66) Areas and Perimeters OF POLYGONS (I)

A **perimeter (P)** is the distance around the polygon. To find it, add the lengths of all sides. **The Area (A)** of Polygons can be found in different ways (See Examples below):

Examples: Find the area and the perimeter for each of the following polygon (Assume all units are in feet).

Area of square $= side^2$
$$A = 5ft \times 5ft = 25ft^2$$

Perimeter (P) $= 5 \times 4 = 20ft$
(Since there are four equal sides)

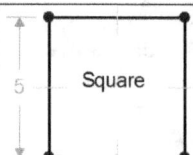

Area of a triangle $= \dfrac{\text{height} \times \text{Base}}{2}$

$$A = \dfrac{3 \times 2}{2} = 3ft^2$$
$$P = 3 + 4 + 2 = 9\ ft$$

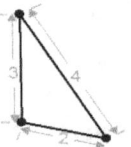

Area of Rectangle = Length $\times$ Width
$$\mathbf{A = L \times W}$$

$$= 7 \times 4 = 28ft^2$$
$$P = 2L + 2W = 2 \times 4 + 2 \times 7$$
$$P = 8 + 14 = 22ft$$

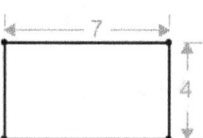

Area of a trapwezoid

$$= \tfrac{1}{2}\text{height (Base1 + Base2)}$$
$$A = \tfrac{1}{2}h\ (b_1 + b_2)$$
$$A = \tfrac{1}{2} \times 4\ (3 + 6) = 18\ ft^2$$
$$P = 3 + 5 + 4 + 6 = 18ft$$

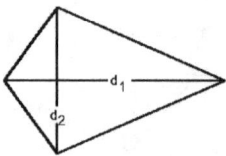

Kite:
A Kite has **two diagonals** d_1 *and* d_2 that are perpendicular with each other.

The area of a kite is the total space inside the boundary of a kite.

Area of the kite $= \dfrac{1}{2}d_1 \times d_2 = \dfrac{1}{2}\ (10) \times (4) = 20$

Assume in this example d1= 10 and d2 = 4

Area and Perimeter Practice

Problems 1 to 9: Find the areas and the perimeters. Unless shown, assume all units are inches:

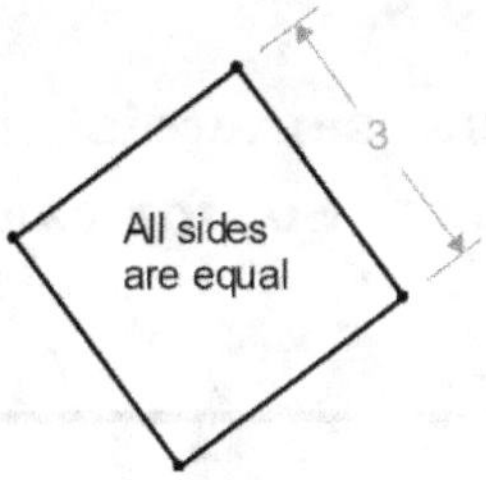

All sides are equal 3

1)

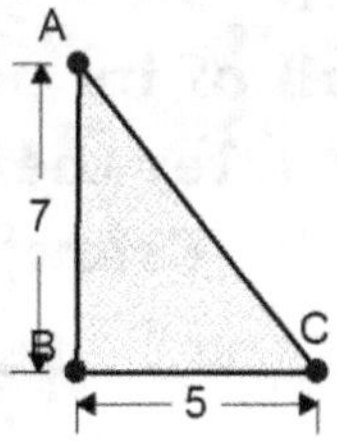

A 7 B 5 C

2)

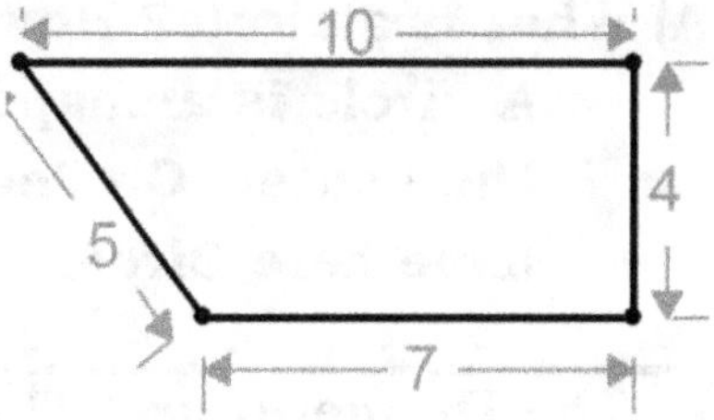

10 4 5 7

3)

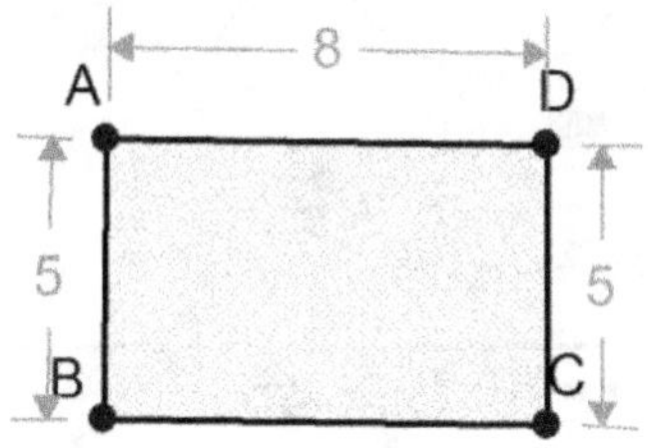

A 8 D 5 5 B C

4)

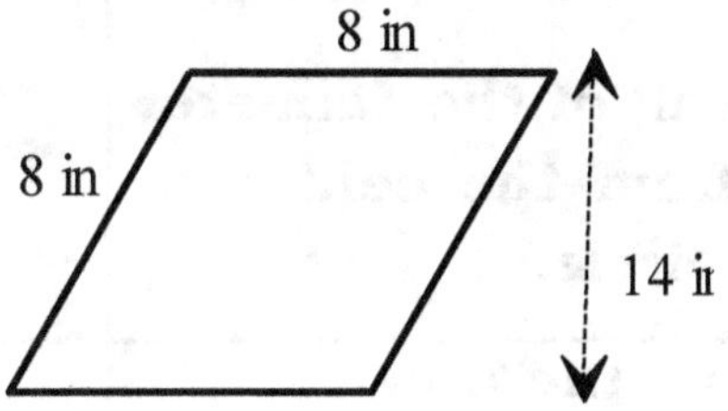

8 in 8 in 14 in

5)

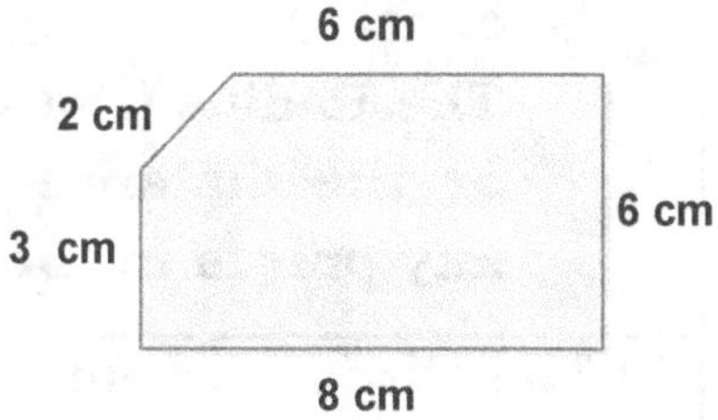

6 cm 2 cm 6 cm 3 cm 8 cm

6)

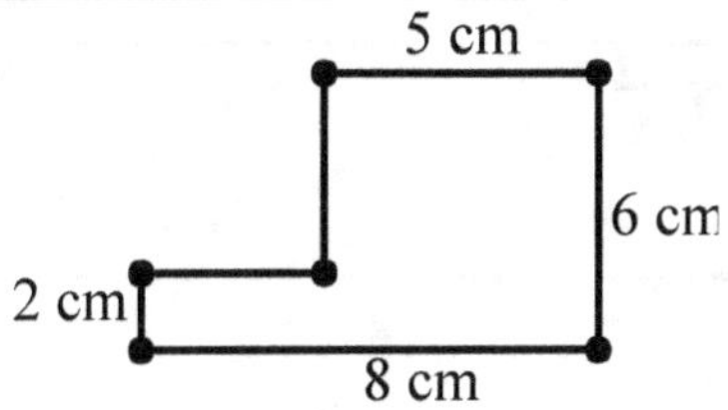

5 cm 6 cm 2 cm 8 cm

7)

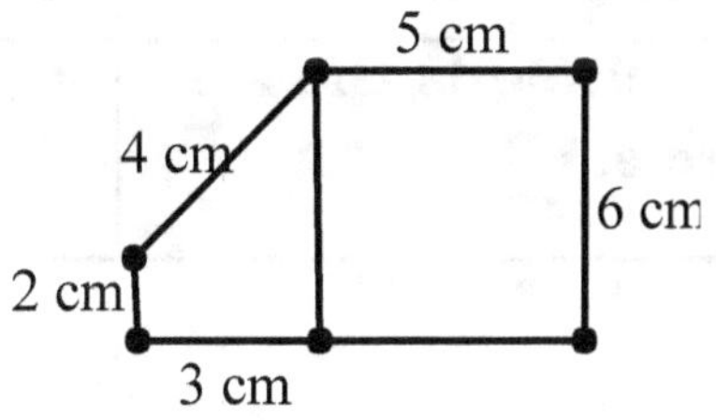

5 cm 4 cm 6 cm 2 cm 3 cm

8)

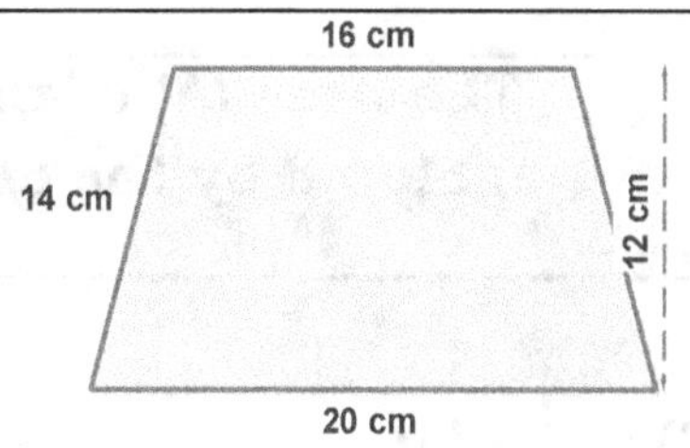

16 cm 14 cm 12 cm 20 cm

9)

The premeter is 18, find the length of the missing side HK?

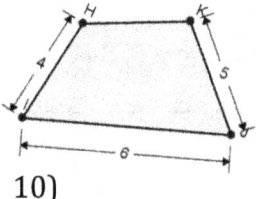

10)

(a) Find the area of triangle JHG
(b) Find the area of the trapezoid.

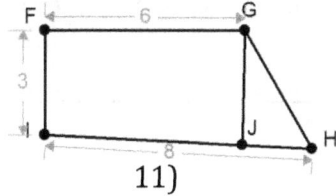

11)

67) Circles: AREAS and Circumference

A) What is a circle? How it is named?

A circle is a shape that has all of its points the same distance from the center. Circles are named after their centers. So, for example we have here Circle A and Circle C. (Below).

B) The Diameter and The Radius

The Diameter (d) connects side to side of the circle passing through the center.

The Radius (r) is half of the diameter or the distance from the center to any points in the circle.

$$r = \frac{d}{2}$$

C) The Circumference and the Area:

The circumference (C) is like the perimeter: It is the distance around the circle.

$C = 2\pi r$ or $C = \pi d$

r is the radius

$\pi = 3.14$

The Area (A) of the circle is the area enclosed by the circle.

$A = \pi r^2$

D) Calculations: Example:

Calculate the diameter, the circumference and the area of a circle with a radius of 6 cm.

$$d = 2r = 2 \times 6 = 12\ cm$$
$$C = \pi d = 3.14 \times 12cm = 37.68cm$$
$$A = \pi r^2 = 3.14 \times (6cm)^2 = 18.84cm^2$$

Circle Practice

Find the circumference, and the area of the following circles:

1) $d = 4cm$ $C =?$ $A =?$

2) $r = 5\ cm$ $C = ?$ $A =?$

3) $d = 14\ cm$ $C =?$ $A =?$

4) $r = 7cm$ $C =?$ $A =?$

5) Name the Circles and then find the length of the diameter, the circumference and the area of each circle.

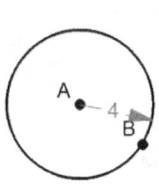 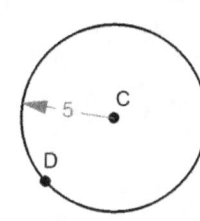 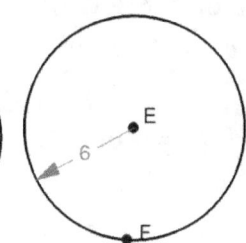

6) **True or False:** A circumference of a circle can be found by multiplying 3.14 with the diameter of the circle.

7) **True or False:** A circumference of a circle is π times its radius?

8) A pizza has a diameter of 8 inches. Find its (a) circumference and

     b) Its area.

68) Area and Perimeter of a kite

1) Find the area of a kite if $d_1 = 10\ cm\ and\ d_2 = 4\ cm$

A kite has **two diagonals** $d_1\ and\ d_2$ that are perpendicular with each other. They bisect at right angles.

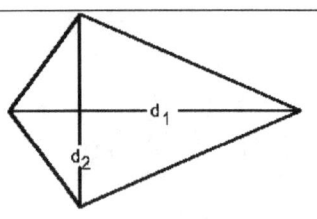

The area of a kite is the total space inside the boundary of a kite.

$$\text{Area of the kite } = \frac{1}{2}d_1 \times d_2$$

$$\text{Area of the kite } = \frac{1}{2}(10) \times (4) = 20$$

Find d_2 if $d_1 = 12\ cm$ and the area is $60cm$

$$\text{Missing diagonal } = \frac{2 \times Area}{known\ diagonal} = \frac{2 \times 60}{12} = 10cm$$

Find the surface area of a kite with the given diagonals.

1) $d_1 = 40\ cm\ and\ d_2 = 10\ cm$

2) $d_1 = 100cm\ and\ d_2 = 20cm$

3) $d_1 = 18\ cm\ and\ d_2 = 14cm$

Find the length of the missing diagonal:.

4) $d_1 = 4\ cm\ and\ Area = 10\ cm$

5) $d_2 = 5\ cm\ and\ Area = 4\ cm$

6) $d_1 = 20\ cm\ and\ Area = 80\ cm$

69) Finding Surface Area and Volume of a Cube

Example: Find (a) the surface area and (b) the volume of the cube that is 8 cm in each side

a) ***Surface Area of a cube with side $S = 6S^2$***

Surface area $= 6 \times 8^2 = 6 \times 64 = 384 cm^3$

b) ***Volume of a cube with side $S = S^3$***

$Volume = 8^3 = 512 cm^3$

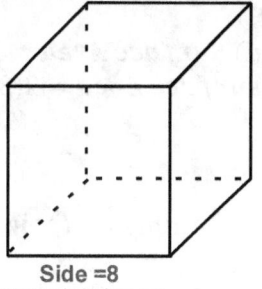

Side =8

Find the surface area of each of the following cubes with the given side. Show your calculations

1) S =5   2) S= 8   3) S= 10

4) S = 6   5) S= 12   6) S= 13

Find the volume of each of the following cubes.

7) S = 10   8) S= 4   9) S= 9

70) Finding Surface Area and Volume of Rectangular Prism

Example : Find (a) the surface area and (b) the volume of a rectangular prism with the height of 3ft., length of 6 ft., and width of 4 ft.

a) $Surface\ Area = 2(h \times l + h \times w + l \times w)$
$surface\ Area = 2(3 \times 6 + 3 \times 4 + 6 \times 4)$
$$= 2(18 + 12 + 24)$$
$$= 2(54) = 108\ in^2$$

b) $Volume = l \times w \times h$
$$Volume = 6ft \times 4ft \times 3ft.$$
$$= 72ft^3$$

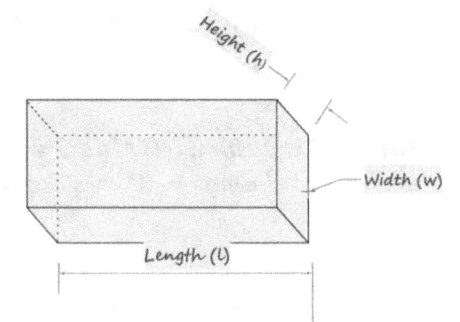

Find the surface area of a rectangular prism with the following dimensions. Make sure to round it to the nearest whole number.

1) Height = 6ft; length= 4ft; width = 5 ft.

2) Height = 4ft; length= 7ft; width = 3 ft.

3) Height =10ft; length =3ft; width = 2 ft.

4) Height = 12ft; length= 5ft; width = 5 ft.

Find the volume of a rectangular prism with the following dimensions:

5) Height = 6 cm; length= 4 cm; width = 5cm

6) Height = 12 cm; length= 10 cm; width = 8 cm

7) Height = 9 cm; length= 3 cm; width = 12cm

8) Height = 11 cm; length= 9cm; width = 10 cm

71) Finding Surface Area of Sphere

a) **Surface Area** $= 4\pi r^2$
$$= 4 \times 3.14 \times 5^2 = 314 cm^2$$

b) **Volume** $= \dfrac{4}{3} \pi r^3 =$
$$= \dfrac{4}{3} \times 3.14 \times 5^3 = 523.3\ cm^3$$

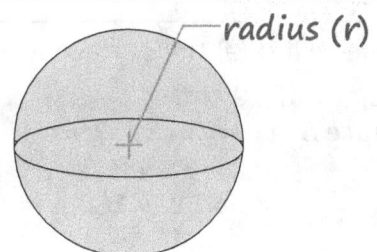
radius (r)

Find the surface area of each of the following sphere with the given radius in cm. Round your answer to the nearest whole number.

1) r = 3

2) r = 8

3) r = 10

4) r = 7

5) r = 4

6) r = 12

Find the volume of each of the following spheres with the given radius. Show your calculations

7) r = 2

8) r = 7

9) r = 5

72) Find Surface Area and Volume of a Cylinder

Example : Find (a) the surface area and (b) the volume of a **cylinder** with radius of 3 cm and height of 10 cm.

a) **Surface Area** $= 2\pi r^2 + 2\pi rh$

$= 2 \times 3.14 \times 3^2 + 2 \times 3.14 \times 3 \times 10$

$= 56.52 + 188.4 = 244.92 cm^2$

b) **Volume** $= \pi r^2 h = 3.14 \times 3^2 \times 10 = 282.6 \ cm^3$

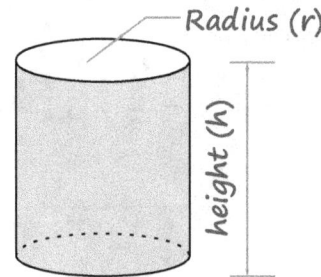

Find the surface area of each of the following cylinder with the given radius and height. The dimensions are all in meters (m). Round your answer to the nearest whole number.

1) r = 3; h = 8

2) r = 4; h = 10

3) r = 10; h = 20

4) r = 7; h = 12

5) r = 11; h = 21

6) r = 12; h = 16

Find the volume of each of the following cylinder with the given radius and height. The dimensions are all in meters (m). Round your answer to the nearest whole number.

7) r = 2 ; h = 8

8) r = 6; h = 30

9) r = 8; h = 12

73) Find Surface Area and Volume of a Cone

Example: Find (a) the surface area and (b) the volume of a cone with slant length of 10 ft. , height of 6 ft. and radius of 5 ft.

a) **Surface Area** $= \pi rs + \pi r^2$

$\qquad = 3.14 \times 5 \times 10 + 3.14 \times 5^2$

$\qquad = 157 + 78.5$

$\qquad = 235.5 \, cm^2$

b) **Volume** $= \dfrac{1}{3} \pi r^2 h$

$\qquad = \dfrac{1}{3} \times 3.14 \times 5^2 \times 6$

$\qquad = 157$

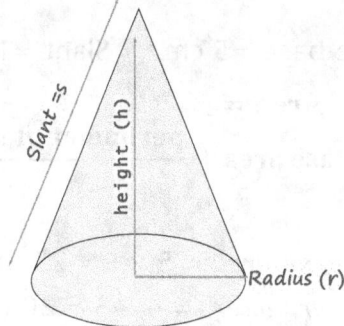

Find the surface area of a cone with the given dimensions. All dimensions are in ft.

1) r = 3; s = 8   2) r = 4; s = 10   3) r = 5; s = 20

4) r = 7; s = 12   5) r = 6; s = 21   6) r = 8 ; s = 16

Find the volume of a cone with:

7) r = 2 ; h= 8;   8) r = 6; h= 30   9) r = 8; h= 2

74) Finding Surface Area and Volume of Pyramid

Find the surface area and volume of the pyramid shown:

Assume base =5 cm ;   Slant =10 cm;   Height= 8 cm

Surface Area =

$$\textbf{Base area} + \frac{\textbf{perimeter of base} \times \textbf{slant height}}{2}$$

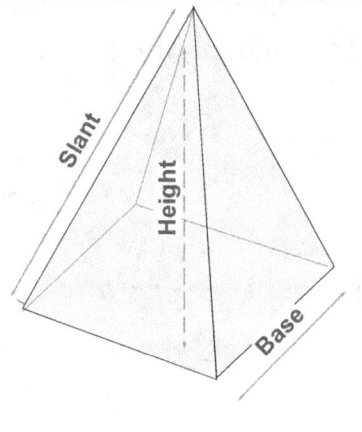

$base\ area = 5 \times 5 = 25$

$Perimeter = 4 \times 5 = 20$ (four sided)

$$Surface\ Area = 25 + \frac{20 \times 10}{2} = 125\ cm^2$$

b) $Volume = \dfrac{1}{3} \times base\ area \times height$

$$= \tfrac{1}{3} \times 25 \times 8 = 66.7 cm^3$$

Find the surface area of the square pyramid with the given slant and base. Use 3.14 for π.

1) base = 4 cm;   slant = 10 cm

2) base = 3 cm;   slant = 9 cm

3) base = 4 cm;   slant = 9 cm

Find the Volume of the square pyramid with the given slant, base and height. Use 3.14 for π.

4) base = 5 cm;   height =8 cm.

5) base = 3 cm;   height =8 cm.

6) base = 4 cm;   slant = 9 cm;   height =8 cm.

75) Review of three Dimensional Area and Volumes

Find the surface area of the following figures with the given dimensions:

1) **Cube**: Side length = 15 cm.

2) **Rectangular prism**: height = 12ft; length = 6ft; width = 4ft

3) **Sphere**: radius of 11 cm.

4) **Cylinder**: radius of 10 m and a height of 15 m.

5) **Cone** with a radius of 10 cm and slant of 14 cm.

6) **Square pyramid** with a slant of 12 m and a base of 5m.

Find the Volume of the following figures with the given dimensions:

7) **Cube**: Side length = 15 cm.

8) **Rectangular prism**: height = 12ft; length = 6ft; width = 4ft

9) **Sphere**: radius of 10 cm.

10) **Cylinder**: radius of 12 m and a height of 15 m.

11) **Cone** with a radius of 9 cm, slant of 14 cm and a height of 11 cm.

12) **Square pyramid** with a slant of 11m and a base of 5m, and a height of 10 cm.

You have completed the course!

Congratualtios for your hard work!

SELECTED ANSWERS

1) Integer Situations/ Evaluations **1) negative   3) Positive 5) Positive 7)   Negative**
   9) 20   11) 17   13) 22   15) 11

Extra Practice: Evaluate the expressions

1) 7   3) 0   5) 1   7) 21 9) 15 11) 24 13) 15 15) 20   17) 14

2) **A) 1) <   5) < 7) >   B) 5)) $-11, -10, -5, -4, -1 \, |-7|$**

3) 1)-10 3)4 5)-9 7)6 9)4 11)-5 13)-37 15)-21 17)-24 19)5

4) 1)-13 3)11 5)13 7)-26 9)-26 11) -15 13)-17 15)-17 17)-8 19)-3

5) 1)-4 3)12 5)-17 7)-36 9)12 11)23 13)-19 15)-13 17)-6 19)-75

6) 1)17 3)-38 5)-24 7)-21 9)90 11)-9 13)7 15)-6 17)-57 19)--10

7) 1)12 3)-32 5)-52 7)30 9)4 11)-9 13)12 15)28 17)-320 19)

8) 1)2 3)-2 5)3 7)-6 9)1 11)-9 13)-5 15)9 17)2 19)-1

9) 1)-10X 3)8x+4y 5)-2a-4b 7)-17m 9)-8x-9 11)5y 13)-17t 15)-21t 17)-5x 19)-7y

10) 1)24 3)12yz 5)8ab 7)-24xy 9)3xy 11)24x 13)0 15)-48x 17)16w 19)24w

11) 1)-4 3)6 5)-10 7)-18 9)7 11)15 13)-46 15)-16 17)30 19)-21

12) 1)96 3)72 5)6 7)48 9)-24 11) 13)12 15)216 17)-32 19)60

13) 1)15 3)7 5)18 7)9 9)18 11)12 13)-4 15)14 17) a) x= 2+2y b)14 18)b) 75

14) 1)20 3)22 5)7 7)15 9)1 11)5 13)12 15) 3

15) 1)6 3)-3 5)7 7)-6 9)0 11)-9 13)4 15)-20 17)10 19)9

16) 1)14 3)2 5)15 7)-4 9)22 11)7 13)-4 15)6 17)20 19)-9

17) 1)-4 3)-11 5)7 7)-4 9)-4 11)7 13)-4 15)-20 17)4 19)1

18) 1)5 3)-5 5)6 7)8 9)7/2 11)-2 13)-2 15)1 17)-9 19)-1/2

19) 1)10 3)-96 5)36 7)36 9)14 11)16 13)-32 15)-120 17)10 19)0

20) 1)4 3)14 5)-9 7)1 9)55 11)0 13)-16 15)-5 17)51 19)-5/7

21) 1)1/5 3)4/7 5)4/3 7)-7/40 9)31/7 11)-1/4

22) 1)1.8 3)-1.4 5)0.5 7)-10 9)-12 11)-0.7 13)0.1 15)-500 17) 19)-17

23) 1)-7 3)18 5)-96 7)-5 9)-65 11)0 13)-24 15)1/3 17)-22 19)10/9

24) 1)30-15 3)20×2 5)15/7 7)8-7 9)2 11)the quotient of 5 and 9

25) 1)x+11 3)20+x 5)5x-15 7)7/x 9)3×2+9 11)2x+2x

26) 1)x=7 3)13 5)x=52 7)x=3 9)x=9

27) 1)x=7 3)2 5)-9 7)2 9)1 11)-2 13)2 15)-1

28) 1)-20 3)16 5)-42 7)28 9)6 11)-36 13)50 15)-6

29) 1)-6x+24 3)-3z+12y 5)-4a+12b 7)-6x+12y 9)-3x+3y 11)15-12x 13)-4x-4y 15)10y+15

30) 1)9x-21 3)-4y 5)-3 7)21x+6 9)x-y 11)-4 13)2x-10 15)4y 17)-6a+6b 19) $10y^2 - 2y$

31) 1)2x 3)7x/9 5)6y/35 7)5t/7 9)9t/10 11)-17m/12 13)2t/3 15)0 17) 19)

32) 1)2 3)-3 5)-4 7)-1/2 9)-1 11)-1 13)-3 15)-1 17)1 19)-1/3

33) 1) 7.5 3)2 5)9/13 7)-3 9)-4 11)-4/3 13)9/2

34) 1)14 3)11 5)1/3 7)-5/2 9)-7/5 11)21 13)-1 15)7/4

35) 1)12 2)56 3)125 4)16 5)24 6)4.9 7) 216

37) 1) 3)$x \leq 5$ 4)$-2 < x \leq 3$ 5)$-2 \leq x \leq 0$ 7) x<10 9)-3<x<3

38) 1) $13x^2$ 3) $2x^3$ 5) $7y^2 + 2x^2$

39) 1)1 3)124 5)19 7) $9y^7$ 9)247 11) $3y^4$ 13) $-7y^4$ 15) $4y^3 + 19y^2$

SELECTED ANSWER CONTINUED

40) 1) 3^8 3)49   5) $12^3 = 1728$ 7) y^4 9)15   11)0

41) 1) x^9 3) y^6 5)1   7) y^2 9) x^{15} 11) 81

42) 1) $1/x^3$ 3)   5) $1/x^6$ 7) $1/x^3$ 9) $1/x^5$ 11)1

43) 1)81   3)64   5)256   7)12   9)648

44) 1)3520   3)0.000561   5)802   7)0.61   9)650000   11)821300

45) 1) 4.5×10^{-2} 3) 4.1×10^4 5) 2.1×10^{-3} 7) 6.1×10^5 9) 3.2×10^{-3} 11) 2.1×10^2

47) 1) B(3,-3) J(-3, 0), L(0, -7), P(5,0), z(2,-2)

49) Hawa (-2,4), Jama (1,2) Aisha (3,-1)   3), Nawal (-6,5)

54) 1)A   2)E   3)B   4)C   5)F   6) D

57) 1) 3   2)-2   3)-1   4) 0   5) 0   6)Undefined

58) 1)-1   3)1   5)2/7   7)-9/7   9) 1   11)-2/3   13)-9-   15) -3/2   17)-5   19)-2

59) 1)y=2x+1   5)y=5x   9) y=-3x-1

61) 1) m= 2 Y-int= 3   3) m= -4 Y-int= -2   5) m= 1 Y-int =0

62) 1)10   3)6.7   5)13.2   7)8.1

63) 1)2.8   3)5   5)5   7)12.2   9)18.9

66) 1)9,12   3)34, 26   5)112,32   7)36, 28   9)216, 64

67) 1)C=12.56 A= 12.56   3) C= 43.6   A= 153.9   7) False   8) C= 24.8 , A= 50.24

68) 1)200   3)126   5)8/5

69) 1)150   3) 216   5) 864   7) 1000   9)729

70) 1) 148   3) 112   5) 120   7) 324

SELECTED ANSWER CONTINUED

71) 1) 113.1   3) 1256.6   5) 201   7) 33.5   9) 524

72) 1) 207.3 3) 1885.7 5) 2212.6 7) 100.6   9) 2413.7

73) 1) 103   3) 396.25   5) 508.7   7) 33.5   9) 134

74) 1) 96   3) 88   5) 24

75) 1) 1350   3) 1500   5) $240\pi = 753.6$ 7) 3375   9) $133\pi = 419$ 11) $297\pi = 932.6$